Neuroinflammation in Vascular Dementia

Neuroinflammation in Vascular Dementia

Gary A. Rosenberg, MD
Professor of Neurology
Departments of Neurology, Neurosciences,
Physiology and Cell Biology, Mathematics and Statistics
Director, Center for Memory and Aging University of New Mexico
Health Sciences Center Albuquerque, New Mexico, United States

ELSEVIER
ACADEMIC PRESS
An imprint of Elsevier

Academic Press is an imprint of Elsevier
125 London Wall, London EC2Y 5AS, United Kingdom
525 B Street, Suite 1650, San Diego, CA 92101, United States
50 Hampshire Street, 5th Floor, Cambridge, MA 02139, United States
The Boulevard, Langford Lane, Kidlington, Oxford OX5 1GB, United Kingdom

ISBN: 978-0-12-823455-6

For Information on all Academic Press publications
visit our website at https://www.elsevier.com/books-and-journals

Publisher: Nikki P. Levy
Acquisitions Editor: Joslyn T. Chaiprasert-Paguio
Editorial Project Manager: Aleksandra Packowska
Production Project Manager: Omer Mukthar
Cover Designer: Miles Hitchen

Typeset by MPS Limited, Chennai, India

Contents

Chapter 1

Clinical aspects of diagnosis in dementia

Introduction

The incidence of dementia has increased worldwide, paralleling the increase in the numbers of elderly individuals referred to as a "silver tsunami." While there are a number of factors involved in this increase in people with dementia, the major one is a demographic shift related to the aging of populations worldwide. Sixty-five has been universally accepted as the age of retirement since the 1800s, when Bismarck (1815–98) set the age to a time beyond the lifespan of the average citizen in Germany so that social benefits would be given to only a few people. However, this age is no longer considered old, and many people are living into their 90s and 100s. Because dementia is a disease of old age, there is a dramatic increase in the number of people with cognitive problems. The main driver of this increase is the aging of populations in both the developed and developing countries. The percentage of people living to be over 100 is high in a number of countries, particularly in Japan, Italy, and France (Fig. 1.1). Senescence is the major factor involved in the incidence of dementia with the chances of being demented increasing drastically after age 80. While age is not a risk factor that can be avoided, there are ways to slow the progress of dementia related to changes in lifestyle; slowing of the age of onset of dementia is possible with a reduction in vascular risk factors, primarily lowering blood pressure, treating diabetes, reducing lipids, controlling sleep apnea, and increasing exercise: this along with adherence to a healthy diet and regular exercise has been proven to slow the onset of cognitive decline.

The increase in elderly, particularly in developed countries, has raised concerns that the cost of caring for those with cognitive impairment will overwhelm the available resources and that many countries will be unable to cope with this marked increase in medical and social services required. In spite of an influx of younger individuals through immigration, which has slowed the aging of the population in the United States, the government has become alarmed and has greatly increased the funds allocated through the National Institutes of Health (NIH) for research into treatments for Alzheimer's disease (AD) and related dementia. As treatments have

Neuroinflammation in Vascular Dementia. DOI: https://doi.org/10.1016/B978-0-12-823455-6.00009-2

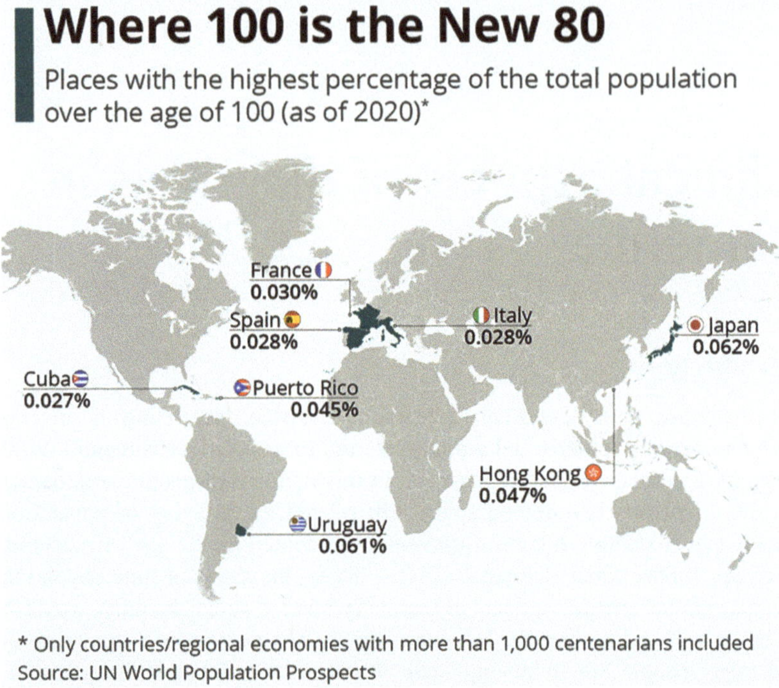

FIGURE 1.1 Places with the highest percentage of the total population over the age of 100 as of 2020. Only countries with more than 1000 centenarians are included. *Countries where 100 is the new 80 permission requested from UN 11–24–21.*

improved for medical conditions, primarily heart disease and cancer, there has been an increase in the number of people developing cognitive impairment. It is currently estimated that there are 6 million people with dementia and that number will double by 2050. Due to the reduced birth rate and increased longevity in developed countries, fewer people will be available to care for those with dementia, compounding the problem will be a reduced number of geriatricians. To meet these needs, the United States government has increased spending to expand planning and to implement plans for long-term care and research into diseases of the elderly. While the focus of this effort has been primarily on AD, there has been a growing awareness of a number of other diseases that are important causes of dementia.

The interest in neurodegenerative diseases of the brain is relatively recent. There was little interest in dementia, with most cognitive loss thought to be due to "hardening of the arteries," and memory loss was considered part of the aging process. Few autopsies were done on older people leaving uncertain the numbers and causes of dementia. The original case report of a patient with early onset dementia by Alois Alzheimer was over 100 years ago; since autopsies on elderly patients with dementia were uncommon, there was little progress in our understanding of the pathological basis of diseases of the elderly. Two major events caused a paradigm shift in the 1970s: the first was a letter to the editor of a neurological journal by Dr. Robert Katzman expressing concern that many patients with dementia were being missed, and the numbers were much larger than thought. The other event were advances in neuroimaging that started with the discovery of computed tomography (CT) by Dr. Housefield in England, leading to the rapid growth in neuroimaging, first with x-ray-based CT and shortly thereafter with magnetic resonance imaging (MRI). These revolutionary changes in the ability to watch pathology develop during life stimulated a major increase in interest in dementia. Concomitant with the growth of imaging, there were major advances in the understanding of the molecular mechanisms, such as the role of amyloid and other abnormal proteins that accumulated in the brain with aging and were important in the pathological damage. The support of the NIH was crucial in the rapid growth in information that has continued over the past 50 years; the first center devoted to understanding and developing treatments for AD was started in a nursing home in the Bronx, New York, by Dr. Katzman. From that humble start, the number of AD research centers (ADRCs) has grown to 34. These centers are at the forefront in building knowledge of the pathophysiology not only of AD, but of all the causes of dementia in the elderly.

A major achievement of the research into the etiology of dementia over the past 30 years has been the recognition of the role of misfolded proteins that are accumulating in the brain and leading to cell death: the two main proteins are amyloid and tau (Long & Holtzman, 2019). Following that discovery, a number of treatments were developed to remove these abnormal proteins. In spite of intensive study with these promising agents, the results have been disappointing, and no disease modifying agent has been discovered although several are being tested. As a result of this failure to find a cure, a number of new research directions are now being explored. One of the most important is due to the recognition that abnormal proteins, many of which are misfolded proteins, stimulate an inflammatory response; this sterile inflammation is due to innate immunity. Other proteins besides amyloid and tau can accumulate in the brain, causing inflammation, and leading to cell damage. These proteins are found in forms of dementia that are less prevalent, including frontotemporal dementia (FTD), Parkinson's disease with dementia, and the recently discovered transactive response DNA-binding protein 43 (TDP-43).

In addition to the deposition of abnormal proteins in the brain, aging results in an increase in damage to blood vessels related to vascular risk factors. This increase in vascular damage parallels the deposition of abnormal proteins and the combined effect is to accelerate cell death and cognitive loss. Blood vessel injury interacts with the deposition of abnormal proteins to accelerate the cognitive decline by hastening neuronal death (Tomimoto, 2011). Alois Alzheimer identified neuronal death in the cortex, while Otto Binswanger, a neuropathologist who was a colleague of Alzheimer's, described the role of pathological damage to the white matter (WM) as a result of arteriolosclerosis/vascular damage as a cause of dementia. Alzheimer showed damage to neurons through the newly developed silver staining and the accumulation of plaques, while Binswanger showed that WM injury resulted from arteriolosclerosis in the penetrating blood vessels feeding the deeper parts of the brain, depriving them of nutrients. These initial observations at the turn of the century formed the basis for the concepts that are still prevalent today. Interest in diseases of old age that result in impaired cognition waned because of the paucity of neuropathological studies and the misconception that loss of memory was a normal accompaniment of aging or, as it was called at that time, senility, leading to therapeutic nihilism. Another major factor for the lack of advancement in our understanding of dementia was that few people lived past the age of 70.

In the late 1970s, computed axial tomography (CAT or CT) was developed in England, making it possible to view images of the brain during life that previously had required autopsy. The mathematical formulas that were developed to form tomographic images with x-rays were rapidly applied to other imaging modalities, including magnetic resonance (MR) and positron emission tomography (PET). With the ability to see inside the skull for the first time, the need for pathological verification increased. In the 1980s, another major leap occurred in our understanding of dementia as the methods of molecular biology showed the pathways involved in brain pathology and studies in genetics revealed the genes involved in the familiar forms of dementia. This powerful combination of imaging and molecular science produced the diagnostic tools that are now unraveling disease mechanisms. As these tools have become available to the neurologist and can be used in everyday practice, they have not only improved diagnosis but have also guided attempts at treatment.

Determining the etiology of cognitive decline

The term dementia is descriptive rather than diagnostic. This is the case in most circumstances except for its use in AD where it refers to a late stage where severe symptoms are present. The term dementia is derived from the Latin term, *dementare*, which means "to make insane"; it is defined in Thomas (1985) as an irreversible deterioration with absence or reduction of intellectual faculties due to an organic brain disease. The progressive nature of dementia separates it from the static processes, such as mental retardation or the results of brain injuries. While there are significant causes of dementia in children, including some vascular diseases, dementia is mainly a disease of older adults except in hereditary forms of AD. Severe hypertension can begin in mid-life with vascular dementia in later life. In the early stages, the term mild cognitive impairment (MCI) is used. Most people with dementia have the late onset form rather than an hereditary form.

While primary brain disorders are the major causes of dementia, there are a number of other medical conditions that are accompanied by a decline in intellect; these fall into two categories; those that are potentially treatable and those that are irreversible. One of the primary aims of the clinician prior to initiating a search for neurological causes is to eliminate the medical causes. Vascular risk factors are potentially treatable, and it is particularly important to identify them early before serious damage has occurred. The main risk factor is hypertension; the WM is vulnerable to long-standing hypertension. Treating hypertension reduces the risk of stroke and dementia. Until recently, the guidelines were to lower blood pressure to 140/90. A new study, SPRINT, suggests that the guidelines should be 120/80 (Yang & Williamson, 2019). Mid-life hypertension identified in participants in the long-term Framingham study showed that those individuals with a blood pressure of 140/90 had an increased risk of WM damage on MRI later in life, but the impact on cognition was not determined (Maillard et al., 2012). Studies such as these emphasize the importance of early identification and aggressive treatment of vascular diseases.

Diabetes is an example of a treatable cause of dementia that is common, and increasing in prevalence, particularly type 2. Diabetes has been linked to cognitive loss with insulin resistance being suspected as the cause. Insulin is important in a number of pathways of brain metabolism, including processing of amyloid and tau. Insulin resistance in the periphery leads to complex changes in the brain. Insulin crosses the blood–brain barrier, suggesting that insulin treatment may help in dementia. However, a trial of nasally administered insulin failed to improve dementia (Kellar & Craft, 2020). Small vessel disease is a common finding in both diabetes and hypertension. MRI shows brain atrophy and reduced cerebral blood flow in diabetes, paralleling the changes observable in the vasculature of the retina. Biomarkers of inflammation and blood–brain barrier disruption are found in patients with type 2 diabetes. Given that obesity and type 2 diabetes are increasingly prevalent in younger age groups, further work is needed to understand long-term effects on the underlying pathophysiology and how these relate to the risk and clinical trajectory of dementia. Other vascular risk factors are suspected to be implicated in the development of dementia, but the studies to show this are equivocal: these include elevated lipids and obesity. Many medical conditions can produce a clinical syndrome similar to degenerative dementia, but are potentially reversible with appropriate treatment. Frequently, patients with endocrinological and vitamin deficiencies have alterations in thinking; low levels of thyroid hormone and vitamin B_{12} are causes of cognitive loss and should be tested for routinely. Collagen-vascular diseases can impair cognition; primarily lupus erythematosus. There are other medical conditions that interfere with thinking, but these are rarely encountered during a work-up for dementia since they are generally detected by the internist.

Neurodegenerative causes of dementia

Clinical diagnosis of dementia is complicated by the overlapping of the major clinical syndromes associated with dementia. AD and vascular cognitive impairment and dementia (VCID) are the most common forms of dementia and frequently occur together; the brain pathology may differ, but the phenotypes are similar (Fig. 1.2; Rosenberg et al., 2016). Patients with memory complaints seek a neurological consultation, often after an initial work-up by an internist to rule out medical causes. From the neurological examination it is possible to begin the diagnostic process: patients with prominent memory loss without focal neurological findings have a high probability of having a neurodegenerative disease such as AD. When there are focal findings, such as hyperreflexia and gait imbalance, a vascular cause should be suspected. Some patients with memory loss present with features of Parkinson's disease, raising the suspicion of frontotemporal dementia (FTD) or dementia secondary to Parkinson's disease. When there is a personality change or symptoms related to a psychiatric disorder, FTD is a

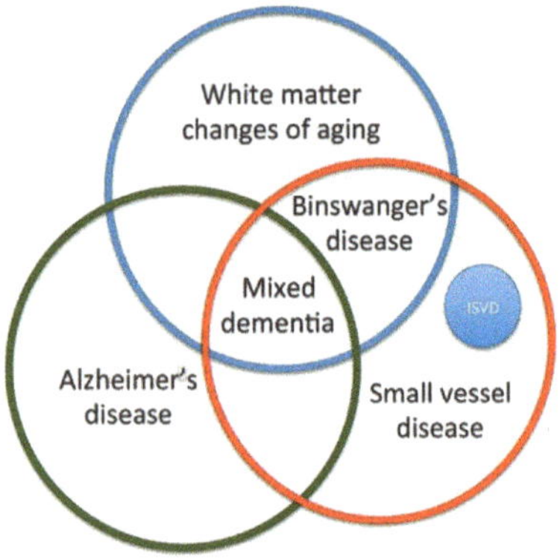

FIGURE 1.2 Overlapping dementia diagnoses. Venn diagram of the various categories of dementia and normal aging. SVD is a major cause of VCI. White-matter changes of aging or leukoaraiosis indicate that the etiology is uncertain. White-matter changes of aging can overlap with SVD in patients with Binswanger's disease. When Alzheimer's disease overlaps with SVD, mixed dementia is the appropriate term. ISVD, such as CADASIL, are a separate category of SVD. *CADASIL*, Cerebral autosomal dominant arteriopathy with subcortical infarcts and leukoencephalopathy; *ISVD*, inherited forms of small vessel disease; *SVD*, small vessel disease; *VCI*, vascular cognitive impairment. *From Rosenberg, G. A., Wallin, A., Wardlaw, J. M., Markus, H. S., Montaner, J., Wolfson, L., Iadecola, C., Zlokovic, B. V., Joutel, A., Dichgans, M., Duering, M., Schmidt, R., Korczyn, A. D., Grinberg, L. T., Chui, H. C. & Hachinski, V. (2015). Consensus statement for diagnosis of subcortical small vessel disease.* Journal of Cerebral Blood Flow & Metabolism. *https://doi.org/10.1038/jcbfm.2015.172.*

diagnostic possibility. After reviewing the blood tests done by the internist prior to the visit, there are other tests that can be ordered: tests for syphilis, human immunodeficiency viruses (HIV), Lyme disease, and coronavirus (COVID-19) are indicated to eliminate infectious causes of intellectual decline.

The Mini-Mental Status Exam (MMSE) or the Montreal Cognitive Assessment (MoCA) are used in screening for cognitive loss; both are excellent to screen for memory problems, but the MoCA is more general and includes tests of executive function, making it the preferred initial test. If it is determined that there is a cognitive deficit, there are several choices for further work-up. A full neuropsychological test can be obtained generally by a clinical neuropsychologist; they can more precisely define the components of the cognitive loss and are often able to separate psychiatric aspects, particularly in those with a suspected underlying depression. When a psychiatric condition is suspected, referral to a specialist for further work-up may be needed.

Misfolded proteins are present in the brains of dementia patients. Amyloid and tau are prominent in AD, tau protein in FTD, corticobasal disease, and traumatic encephalopathy. α-synuclein in Parkinson's dementia, and TDP-43 in hippocampal sclerosis (Box 1.1).

The attempt to link abnormal deposits of amyloid with AD has led to a number of studies to remove the amyloid. To date, all of these studies have failed, bringing the amyloid hypothesis into question. Parallel with the clinical trial failures that have cast doubt on the amyloid hypothesis has been the growth in studies suggesting that multiple factors have stimulated the immune system to mount an inflammatory response. This has opened a search for agents that block inflammation in the brain, and has led to a number of studies. As the abnormal proteins accumulate in the brain, a series of

BOX 1.1

It is common to find multiple pathological changes at the time of autopsy. Elderly patients have four major misfolded proteins that contribute to dementia: amyloid-β (Aβ), phosphorylated tau (pTau), α-synuclein, and transactive response DNA-binding protein 43 (TDP-43). Those with quadruple misfolded proteins have the highest likelihood of being demented. Autopsy studies show that 20% of patients had quadruple misfolded proteins; they had the lowest scores on cognitive tests; about 40% had three abnormal proteins. The presence of α-synuclein and TDP-43 in patients with significant amounts of pTau and Aβ pathology markedly increased the chances of being demented. The presence of co-morbid synuclein and TDP-43 with Alzheimer's pathology complicates the search for therapies (Karanth et al., 2020).

events occur related to stimulation of the brain's innate immunity that involves the resident inflammatory cells, the microglia, which produce inflammatory molecules. The influx of macrophages and T cells further aggravates the inflammation. Since amyloid and tau activate microglia and recruit macrophages that contribute to the inflammatory response, it is possible that the inflammatory response is being driven by these abnormal proteins. Another important factor is the need to remove the abnormal proteins, which is done by their removal into the interstitial and cerebrospinal fluids; these act as the brain lymphatic system; impairment of removal is another way that the accumulated proteins can continue to cause damage to brain cells (Benveniste & Nedergaard, 2021). While misfolded protein has a deleterious effect when present alone, the presence of multiple abnormal proteins most likely accelerates the inflammatory response. Other factors, such as APOEε4 and vascular disease, further aggravate the inflammatory response. The recognition of the cumulative effect of multiple pathological processes stimulated research into alternative theories of AD when the amyloid hypothesis was found to be only partially correct. This search for alternative mechanisms has lead to the realization of the important role of vascular disease and the emphasis on lifestyle as a way to slow the progress of both AD and vascular dementia.

Some form of neuroimaging is essential at an early stage of the work-up. Depending on the type of brain scanner available, this can be either a CT scan or a MRI. While the CT provides useful information about masses such as brain tumors, hydrocephalus or subdural hematomas, the poor resolution makes diagnosis of diseases primarily involving the gray or WM difficult. Because of its higher spatial resolution and ability to visualize injury to the WM, MRI is optimal; it is able to provide all the information contained in a CT scan as well as a large amount of information on the nature of the underlying disease process, which is not available from CT. The MRI shows atrophy of the gray matter in the cortex and hippocampus that is seen mainly in AD. Injury to the WM can be readily seen on the MRI. Normal pressure hydrocephalus (NPH) is easily observed by MRI as well as subdural hematoma, which can be missed on CT, particularly when it is close to the bone. But the major improvement of MRI over CT is in the diagnosis of WM disease. A number of sequences clearly show changes in the WM. Fluid attenuated inversion recovery (FLAIR) is routinely used to identify injury to the WM. However, it also shows the WM hyperintensities that are the hallmark of vascular dementia in many elderly people. The overlap between gray and WM injury signals in AD and VCID and the presence of similar changes in normal elderly has confounded diagnosis and has led to the development of further refinements of MRI, such as diffusion tensor imaging (DTI), which improves diagnostic accuracy. Since WM hyperintensities are often present in patients with AD and always in patients with VCID, separating them from the WM changes of aging, can be problematic. At the interface of both AD and VCID are WM changes of aging and to separate patients from normals requires DTI, which has the advantage of being a better indicator of WM injury and is available on all MRI scanners. When small vessel disease overlaps with AD, the patients are called multiple etiology or mixed dementia.

Alzheimer's disease

Progressive memory loss is most commonly caused by AD, but this common complaint has not been useful by itself for making the diagnosis of AD. Since few papers were published on AD prior to 1970, and the diagnosis was made infrequently, there was little need for diagnostic criteria. However, this changed in the 1970s as more patients were suspected of having AD, and there were no criteria on which to base a diagnosis. A working group began to formulate diagnostic criteria that were published in 1984 (McKhann et al., 1984). These initial attempts at establishing diagnoses used clinical criteria without the use of biomarkers, which later became increasingly important. The initial clinical criteria were: (1) an insidious onset of cognitive impairment over months to years; (2) a clear-cut history of progression of cognitive decline obtained from an informant and confirmed by neuropsychological testing; (3) an amnestic presentation with involvement of episodic memory and poor retention of information; and (4) the lack of other possible diagnoses, such as medical conditions and vascular disease. In 2011, these criteria were further refined: the new criteria maintained the prior clinical criteria, but added several new concepts, including mild cognitive impairment (MCI) that included concern by the patient about a change in cognition, cognitive decline in one or more areas that exceeded expectations for age, continued independent living, and absence of dementia (McKhann et al., 2011). These criteria based on the clinical evaluation included neuropsychological test results. Although evidence was accumulating to support a biomarker approach including cerebrospinal fluid (CSF), MRI, and positron emission tomography (PET) biomarkers, these had not reached the level of acceptance to be included as components of the diagnostic criteria.

A major shift in thinking about AD occurred in 2018 with the publication of "biological" criteria for the diagnosis of AD based on the increased confidence in the use of multimodal biomarkers (Jack et al., 2018). Advances made during the 10 years between the clinical paper and the biological one in detection of amyloid in the CSF, neuroimaging including PET scanning with amyloid ligands, and genetic studies to identify the hereditary forms of AD at early onset facilitated the diagnosis. The biological criteria were based on the detection of amyloidβ and phosphorylated Tau_{181}(pTau) in CSF, which can appear as early as 20 years before onset of symptoms in hereditary AD. There were also advances in MRI that allowed for detection of atrophy of the hippocampus and cortical tissues. Finally, improvement in PET ligands improved detection of amyloid and tau proteins in the brain.

Recently, the demonstration of increased blood–brain barrier (BBB) permeability localized to the subregions of the hippocampus associated with cell loss in AD provided evidence that the BBB injury may contribute to the etiology of AD. To make a biological diagnosis of AD it is necessary to have either CSF demonstrating a reduction in amyloidβ, an elevation in CSF pTau, or a positive PET scan using either amyloid or tau ligands. Another factor is the presence of hippocampal atrophy on MRI. They proposed the formula, ATN, with A indicating amyloid, T tau, and N neurodegeneration as shown by hippocampal atrophy. This is primarily for use in research studies, since obtaining CSF by lumbar puncture is invasive and PET studies are expensive and mainly available at research centers.

Other factors are found in the brains of patients with dementia that need to be included in the ATN formula: the main factor is hypertensive disease. These factors can be included in the ATN formula to take into account other types of pathology. At this time, the MRI shows WM injury with DTI. When the WM is damaged, the formula can be expanded to include a vascular factor, V. Other indicators of vascular disease could also be used. When CSF or PET findings indicate AD proteins are present and the MRI shows WM injury, a diagnosis of mixed dementia can be made. This can be shown graphically by use of a drawing with AD proteins to create an AD biomarker score on one axis and a WM injury score on the other. In this graph, the upper left quadrant with only AD proteins indicates "pure" AD, the lower right quadrant with only WM damage contains patients with SIVD. When neither are present as in the lower left quadrant, the patients are normal or can be labeled as leukoaraiosis, indicating an abnormal FLAIR image with normal DTI and normal CSF studies. The upper right quadrant indicates mixed dementia.

The gold standard for diagnosis of AD is to detect pathological changes in the brain; these include amyloid plaques and neuronal tangles. The patients with only AD will not show changes in the brain of vascular disease. Pathological confirmation of mixed dementia requires the presence of both AD proteins and vascular disease. The typical pathological findings of AD are amyloid plaques and neuronal tau tangles (Fig. 1.3).

Vascular causes of dementia

Prior to neuroimaging, when few autopsies were performed, death certificates often listed "hardening of the arteries" as the cause of death. An early pathological study proposed that the size of the infarct was related to dementia (Tomlinson et al., 1970). The term multi-infarct dementia (MID) was suggested from studies of autopsies and cerebral blood flow in demented patients with those with neurodegeneration showing normal blood flow and reduced cerebral blood flow in those with vascular disease (Hachinski et al., 1975). When it was possible to visualize the brain by CT and MRI during

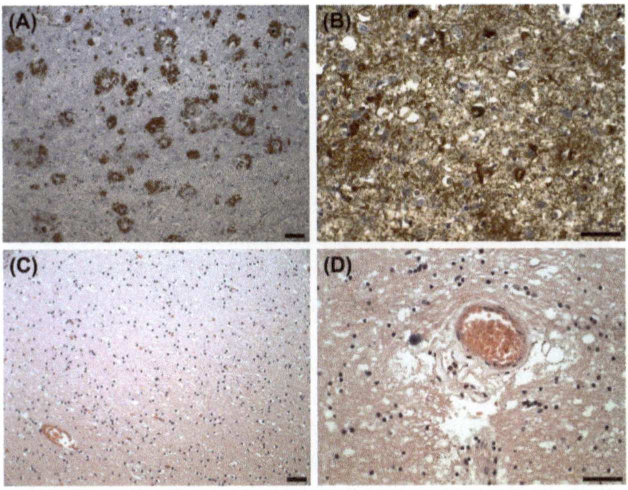

FIGURE 1.3 Pathological findings in patient with AD. Pathological findings in AD patients as seen with light microscopy. (A) β-amyloid plaques in the cortex *(arrow)*. (B) Neuritic tangles *(arrow)* resulting from hyperphosphorylation of tau protein in the cortex. (C) The blood vessels appeared normal in WM hematoxylin and eosin (H&E). (D) H&E staining shows enlarged perivascular space *(arrow)*. *AD*, Alzheimer's disease; *WM*, white matter. *Courtesy of Yi Yang, MD, PhD and Karen SantaCruz, MD.*

life, a group of patients was observed that did not have large strokes, but instead had lacunar strokes or extensive WM disease; it was observed that WM changes were associated with cognitive loss without obvious strokes. New definitions of vascular disease related to dementia began to be used; this resulted in a period of confusion as the optimal descriptive terms were being debated. The umbrella term VCID was selected as the optimal term that incorporates all vascular causes of dementia. VCID can be classified into large and small vessel disease with large vessel involvement resulting in strokes; these are often related to disease of the carotid and the heart with emboli from atherosclerotic plaques and atrial fibrillation. Small vessel disease is commonly the result of vascular risk factors, including hypertension and diabetes, as well as primary diseases of the blood vessels and sleep apnea. The clinical syndromes associated with each of these types of vascular disease differ; large vessel disease causes sporadic strokes, while small vessel disease produces a progressive decline in intellect and gait and balance abnormalities. The progressive nature of small vessel disease makes it more amenable to treatment trials where it is possible to understand the natural history; sporadic disease related to stroke-like events is less predictable and would require much larger numbers of patients for a successful treatment trial.

Neuroimaging methods are excellent for diagnosing both large vessel strokes and progressive WM disease; while CT is useful, MRI is the optimal imaging modality. Clinical findings are less helpful because of the overlapping symptomatology in the diseases causing dementia. Large vessel strokes can generally be diagnosed by the abrupt onset of symptoms and visualization of the large infarcts on imaging. Separating neurodegeneration from small vascular disease is more difficult because both have slowly progressive cognitive impairment with memory more prominent in the former and executive dysfunction more common in the latter. Imbalance, incontinence, and apathy are frequently seen with progressive small vessel disease. Combining these clinical findings with the demonstration of WM hyperintensities on MRI and a history of vascular risk factors provides strong evidence of a small vessel cause of the dementia.

An MRI on a typical patient with VCID lacks obvious strokes and has minimal cerebral atrophy. The MRI of a patient with VCID shows the large areas of WM changes in periventricular and subcortical regions (Fig. 1.4)

Pathological findings in VCID patients include gliosis in the WM, arteriolosclerosis in the deep WM blood vessels occasionally with areas of hemorrhage around damaged blood vessels, referred to as microbleeds, hypertrophy of the fibrotic vessel walls with narrowed vessel lumens, and inflammatory cells found around blood vessels in the WM. The gray matter is normal, and there is an absence of amyloidβ in the cortex and blood vessels, and normal appearing blood vessels in the gray matter (Fig. 1.5).

Mixed dementia—AD and VCID

The frequency of patients with mixed dementia has been shown in autopsy studies to be as high as 70%; however, the ability to make this diagnosis prior to death remains limited. Generally, the MRI is used as a guide to diagnose vascular disease, using different standards. Some studies use FLAIR MRI to identify the WM injury, while in AD both the biological definition based on the finding of AD proteins in CSF or PET studies is used and the clinical diagnosis, relying mainly on neuropsychological findings. This plethora of approaches has hampered the ability to obtain epidemiological data on the incidence of mixed dementia. In addition to an increase in vascular disease in the aging brain, a number of proteinopathies have been identified; these are difficult to diagnose on clinical grounds and are only identified at autopsy. This could explain the poor performance of treatments focused on reducing brain amyloid while leaving a number of other conditions untreated. While the number of abnormal proteins accumulating in the brains of patients during the aging process continues to grow as new pathological entities are described, the relentless increase in the pathology due to vascular disease becomes more important as a natural accompaniment of aging.

Advances in MRI techniques and in biochemical and genetic studies have expanded the tools available to select patients with vascular diseases. Some of these MRI methods have entered routine clinical use with the new

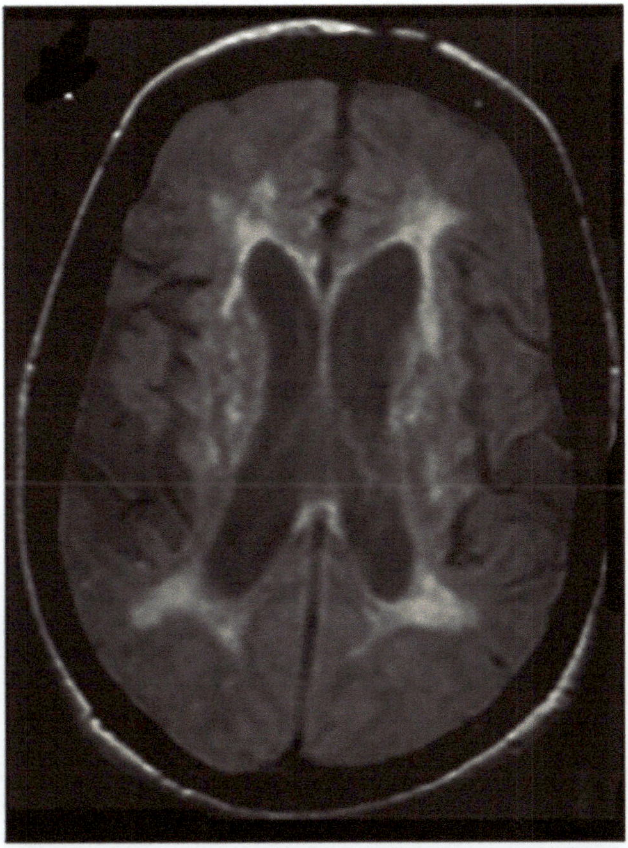

FIGURE 1.4 MRI of patient with small vessel VCID. The FLAIR image shows the large areas of white matter injury (white) in the periventricular and deep white matter. Note the relatively normal ventricular size and the lack of cortical atrophy. *FLAIR*, Fluid attenuated inversion recovery; *MRI*, magnetic resonance imaging; *VCID*, vascular cognitive impairment and dementia.

generation of MRI scanners: DTI provides a more accurate imaging modality than FLAIR for detecting injury to the WM. Newer MRI methods are slowly leaving the realm of research, such as functional MRI with resting state and activation studies, identifying abnormal brain connections due to damage to WM fiber tracts. The first studies to indicate the importance of vascular disease in combination with AD were done by Snowdon and colleagues in subjects that were part of a religious order that agreed to yearly neuropsychological testing with MRI and ultimately to undergo autopsies; this so-called "Nun Study" showed that the presence of AD pathology in the form of amyloid and tau when combined with evidence of vascular disease accelerated the cognitive impairment (Snowdon, 2003). The impact of vascular disease on AD was not appreciated in the large autopsy series collected

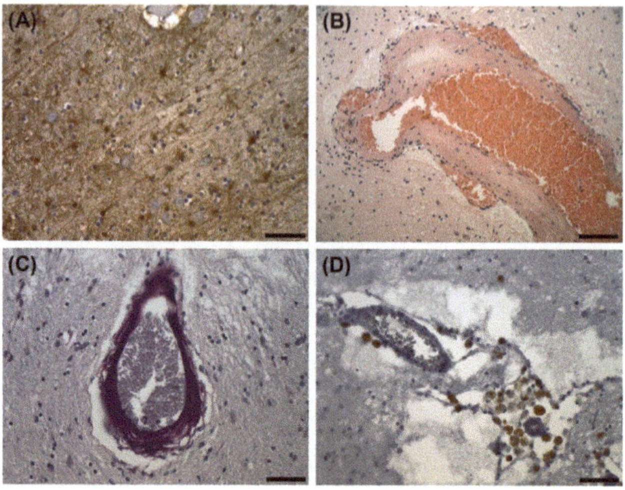

FIGURE 1.5 Pathological findings in a patient with VCID. Pathological findings in VCID patients with light microscopy. (A) Extensive gliosis (GFAP-positive) in the WM. (B) Arteriolosclerosis in the deep WM with areas of hemorrhage around damaged blood vessels. Hematoxylin and eosin(H&E). (C) Hypertrophy of the vessel walls. Periodic acid–Schiff (PAS). (D) Inflammatory cells (CD68) around a blood vessel in the WM. Scale bars = 50 μm in A, B, C, D, and F; 100 μm in E. *VCID*, Vascular cognitive impairment and dementia; *WM*, white matter. *Courtesy of Yi Yang, MD, PhD and Karen SantaCruz, MD.*

over 30 years by the Alzheimer's disease research centers (ADRCs) brain banks, since they had excluded patients with suspected vascular disease in order to obtain a collection of brains with mainly AD. Recently, review of the autopsy material in the ADRCs, revealed an estimated 70% with evidence of both AD and vascular disease. In addition, other proteinopathies were identified, adding to the growing awareness of the importance of multiple pathological changes (Toledo et al., 2013).

Diagnosis of patients with mixed dementia prior to autopsy is now possible by the use of a multimodal approach that utilizes biomarkers; the biological criteria for AD indicate those patients with a high probability of having AD by the finding of low amyloidβ and elevated phosphorylated tau in the CSF or either AD protein in the brain by PET. CSF requires an invasive lumbar puncture while PET is only available in larger centers and is expensive, making it of limited usefulness. Combining MRI measures of white matter injury and AD proteins allows the diagnosis of mixed dementia during life (Caprihan et al., 2021). The ability to detect AD proteins in plasma has made a major impact by allowing diagnosis to be made less expensively and non-invasively. There are several instruments that can detect very small quantities of AD proteins in the blood. As these become more available and accepted in the diagnosis of AD, the detection rate will increase, the time to diagnosis decrease, and the ability to determine treatment effectiveness will improve. This will occur as these ultra-sensitive assay platforms, such as the single molecule assay (SIMOA) system made by the Quanterix Company, enter routine clinical practice. These instruments can measure very small amounts of AD proteins in plasma (Karikari et al., 2020).

Highly sensitive biomarkers for AD and VCID will allow detection of dementia much earlier than is possible clinically. By early detection, disease-modifying therapies can be given at a stage prior to the onset of permanent cell loss. Since these different pathologies have different timing of onset and different rates of progression, the ability to detect the very early stages of these diseases will provide the means for customized treatment strategies, beginning treatment for one before the other; a schematic drawing of this approach is shown in Fig. 1.6 (Rosenberg et al., 2019). Amyloid proteins are accumulated as early as 20 years prior to the onset of symptoms; these are in the amyloid-containing plaques around blood vessels. Detection of the amyloid in the CSF is possible, reflecting the deposition in the brain. Tau proteins are accumulated later than the amyloid, beginning the process of neuronal death. This process of abnormal protein accumulation progresses over many years, reaching a level at some point when an inflammatory process is triggered. A parallel process begins to damage the blood vessels; this can begin in mid-life in individuals with chronic, poorly treated hypertension and diabetes. A small number of patients with CADASIL have a genetic defect in the Notch gene that results in early onset of vascular disease. At some point, depending on a number of environmental and genetic factors, both abnormal AD proteins and damaged blood vessels converge; since each process can cause inflammation, the combination of the two accelerates the damage to the brain. This is manifested by the more severe form of dementia seen in those patients with mixed dementia.

α-Synucleinopathies

Dementia with Lewy bodies

Synucleinopathies (also called α-synucleinopathies) are neurodegenerative diseases characterized by the abnormal accumulation of aggregates of α-synuclein protein in neurons, nerve fibers or glial cells. α-synucleinopathy incorporates clinical entities such as Parkinson's disease (PD), PD with dementia, dementia with Lewy bodies (DLB) and multiple-system atrophy. They are no diagnostic tools to diagnose these diseases during life with a definitive diagnosis done at autopsy. DLB has several clinical presentations, depending on the dominant symptoms: in the early stages of the illness, there are clinical features of Parkinson's disease, MCI, delirium, and psychiatric symptoms. The onset of the illness is insidious and difficult to diagnose. Rapid eye movement sleep behavior disorder (RBD) is found in some patients with DLB, starting many years prior to the full syndrome.

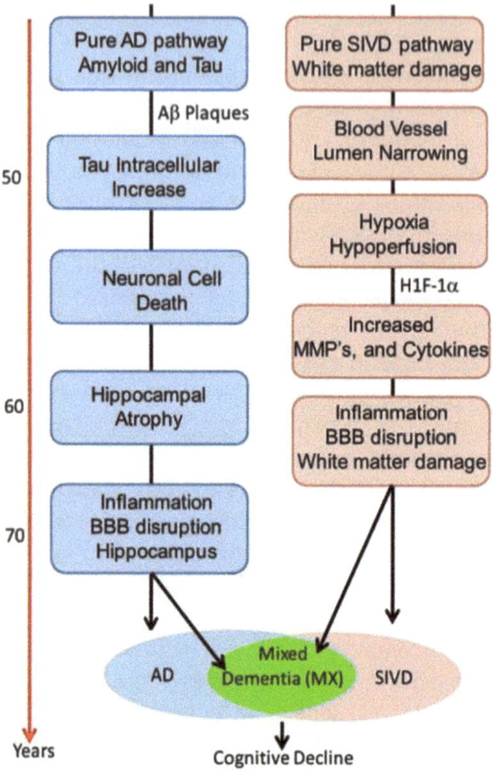

FIGURE 1.6 Dual pathways leading to either AD or VCID. Two possible pathways for progression of pathological changes in the brain. On the left is shown the pathway that starts with deposition of amyloid, which initiates an inflammatory process that ends with the formation of abnormal tau proteins and cell death. On the right, the pathway is shown to be vascular pathology, which also initiates the inflammatory response, but does not necessarily lead to cell death, but extensive injury to the white matter. When both processes merge, there is the creation of the pathology of the mixed dementias. *AD*, Alzheimer's disease; *VCID*, vascular cognitive impairment and dementia. *From Rosenberg, G. A., Prestopnik, J., Knoefel, J., Adair, J. C., Thompson, J., Raja, R., & Caprihan, A. (2019). A multimodal approach to stratification of patients with dementia: Selection of mixed dementia patients prior to autopsy.* Brain Sciences, 9 *(8). https://doi.org/10.3390/brainsci9080187.*

Parkinson's disease is another way that DLB can begin. Fluctuating symptoms are often confounding at onset. Less frequently, DLB can begin as a psychiatric syndrome. There are no defining imaging or fluid biomarkers, although promising ones, such as α-synuclein in the CSF are under development (Kovacs, 2016). A further disorder, multiple system atrophy (MSA), is dominated by glial cytoplasmic inclusions (Papp–Lantos bodies). In addition to these primary synucleinopathies, deposition of α-synuclein is also commonly observed in the brains of individuals with other primary diagnoses.

Aberrant accumulation of α-synuclein is frequently observed in brains with abnormal deposition of pTau, transactive response DNA-binding protein 43-kDa (TDP-43), amyloidβ (Aβ) or prion protein (Kovacs, 2016).

An important role of immune cells as part of the inflammatory response was shown both in human tissue from patients with DLB and in a transgenic mouse with the α-synuclein gene; in one study of T cells and B cells, increased infiltration of CD3 + T cells in the brains of DLB patients was similar to that seen in the α-synuclein overexpressing animal model of DLB. Infiltrating T cells were also found in close proximity to the cell processes of astrocytes and microglia, and accompanied by considerable glial activation and enhanced TNF-α gene expression. In agreement with these findings in α-synuclein transgenic mice, postmortem brain tissues from DLB patients also displayed increased infiltration of CD4 + helper T cells (Iba et al., 2020). In addition to the infiltrating T cells both in the human tissues and in the transgenic mice, there was an increase in microglia and reactive astrocytes (Fig. 1.7). The presence of T cells and activated microglia provides convincing evidence that the α-synuclein pathology is associated with an inflammatory response.

Parkinson's disease with Lewy bodies

PD with dementia (PDD) is a second type of α-synucleinopathy. The clinical and neuropathological features of DLB and PDD are similar, making these a spectrum disorder with differences mainly in the timing of the cognitive decline: in DLB there is relatively early cognitive decline, while it occurs later in PDD (Jellinger & Korczyn, 2018). By the time of death, distinguishing the two diseases pathologically is more difficult. Both include dementia, cognitive fluctuations, and visual hallucinations; both show progressive executive dysfunction, visuospatial abnormalities, and memory disorders. DLB is diagnosed when cognitive impairment precedes parkinsonian motor signs or begins within 1 year from its onset, whereas in PDD, cognitive impairment develops in the setting of well-established PD (McKeith et al., 2017). DLB patients will also develop parkinsonism of increasing severity over the years, although 25% of them never develop parkinsonian symptoms. The overlap of clinical and morphological features has led to the debate of whether DLB and PDD are the same disease, different phenotypic expressions of the same LBD spectrum, or distinct "diseases" (Boeve et al., 2016).

Tauopathies

Frontotemporal dementia variants

Several clinical variants of FTD have been identified. The most common form of FTD is the behavioral variant of FTD, which is characterized by a constellation of behavioral and neuropsychiatric symptoms, including apathy,

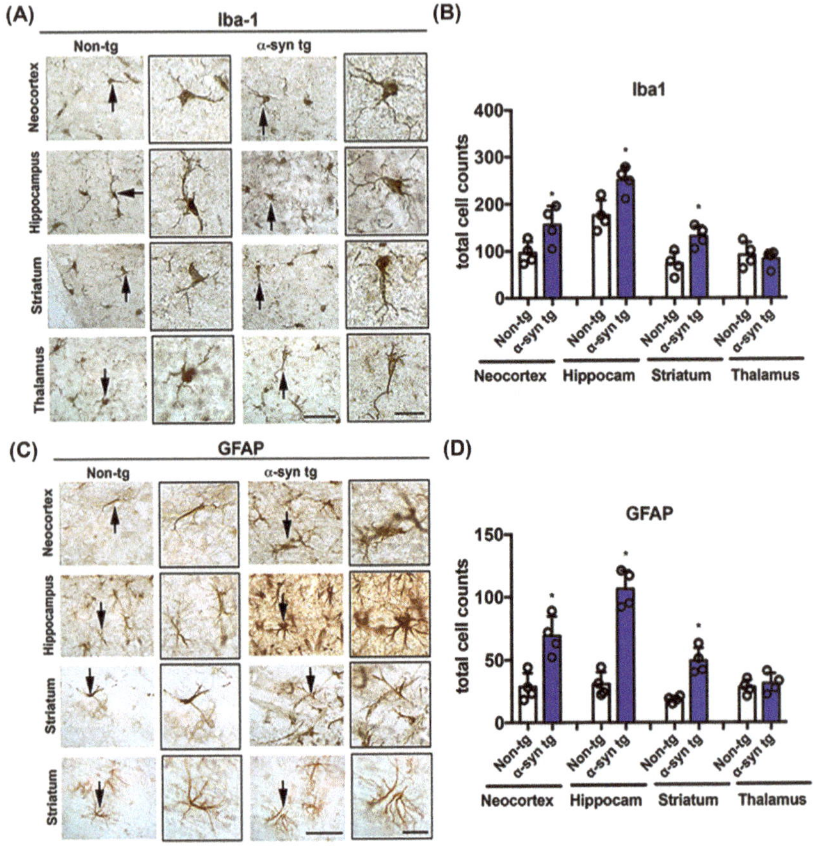

FIGURE 1.7 Immunohistochemical analysis of microglial cells and astrocytes in the brains of α;-synuclein transgenic mice. (A) Representative bright field light microscopy images from the neocortex, hippocampus, striatum, and thalamus of non-tg and α;-syn tg mice immunostained with an antibody against Iba1 (microglial cell marker). (B) Computer-based image analysis showing increased numbers of Iba1 positive cell in the neocortex, hippocampus, and striatum of non-tg and α;-syn tg mice. (C) Representative bright field light microscopy images from the neocortex, hippocampus, striatum, and thalamus of non-tg and α;-syn tg mice immunostained with an antibody against GFAP (astroglial cell marker). (D) Computer-based image analysis showing increased numbers of GFAP positive cells in the neocortex, hippocampus, and striatum of non-tg and α;-syn tg mice. Scale bars = 40 μm (low magnification) and 10 μm (high magnification). Non-tg and α-syn tg mice ($n = 4$). Statistical significance determined by unpaired t-test; $^*P \leq .05$. *From Iba, M., Kim, C., Sallin, M., Kwon, S., Verma, A., Overk, C., Rissman, R. A., Sen, R., Sen, J. M. & Masliah, E. (2020). Neuroinflammation is associated with infiltration of T cells in Lewy body disease and α-synuclein transgenic models.* Journal of Neuroinflammation, 17*(1), 214. https://doi.org/10.1186/s12974-020-01888-0.*

disinhibition in interpersonal interactions, euphoria, depression, emotional withdrawal, alterations in food preferences and eating behavior, irritability, repetitive behavior, and dysexecutive symptoms. Primary progressive aphasia (PPA) is a second form of FTD that refers to three main clinical constructs differentiated by specific types of speech or language deficits: The non-fluent, or agrammatic, variant is characterized by incorrect use of grammar or syntax, labored and/or halting speech with inconsistent sound errors and distortions, and a lack of comprehension of syntactically complex sentences; object knowledge and single-word comprehension are relatively spared. The semantic variant is characterized by semantic aphasia and associative agnosia, disrupted semantic memory with impaired object recognition—anomia and/or single-word comprehension deficits are essential for diagnosis. Finally, the logopenic variant, which is most often a manifestation of underlying AD, involves difficulties with word finding as an early deficit and problems with repetition of sentences (Panza et al., 2020).

Corticobasal degeneration and progressive supranuclear palsy

CBD and PSP are tauopathies. Tau is encoded by the microtubule-associated protein tau gene. Under normal conditions, tau is predominantly expressed in neurons where it facilitates microtubule stability and assembly. There are six isoforms of tau resulting from alternative mRNA splicing of exons 2, 3, and 10. They are divided into 3-repeat-tau or 4-repeat-tau isoforms. The extra repeat is coded by exon 10. Pathological straight tau filaments in PSP and CBD are primarily made up of 4-repeat-tau. PSP and CBD have their own validated neuropathological diagnostic criteria and are considered distinct pathological entities. PSP is characterized by tau-enriched tufted astrocytes and neurofibrillary tangles in subcortical nuclei. The minimal pathologic features for CBD are cortical and striatal tau-positive neuronal and glial lesions, especially astrocytic plaques and thread-like lesions both in white and gray matter, along with neuronal loss in focal cortical regions and in the substantia nigra. The pathological hallmarks for PSP and CBD, tufted astrocytes vs astrocytic plaques, appear not to coexist in the same brain. Accordingly, Western blots of insoluble filamentous tau extracted from affected brain regions of PSP and CBD show different lower molecular weight bands with a 33-kDa band in PSP and two closely related bands of approximately 37-kDa in CBD.

Less than half of the patients with CBD are correctly diagnosed and there is phenomenological overlap with several neurodegenerative disorders. Pathological examinations suggest that CBD begins as an astrogliopathy with tau-positive astrocytic plaques and surrounding threads, but neuronal lesions gradually develop over time. This progression may depend upon intracellular propagation of specific tau strains. Imaging modalities can provide aid in the differential diagnosis: Tau PET ligands remain to be shown to aid in CBD diagnosis. CSF levels of neurofilament light (NfL) are used to distinguish atypical parkinsonian syndromes, including CBD, from Parkinson's disease and controls. However, at the moment, there is no CSF biomarker that can differentiate CBD from PSP and MSA. There is a rapid development in sensitive and large-scale approaches to detect analytes in biofluids. Despite the heterogeneous clinical presentations of CBD, it is important to carefully follow and biosample CBD patients longitudinally and confirm CBD diagnosis postmortem.

Propagation of tau disease reminiscent of CBD has been recapitulated by intracerebral injection of brain homogenates from CBD patients into humanized tau mutant mice. Moreover, isolated aggregated "tau strains" from CBD patients intracerebrally injected into humanized tau mutant mice can cause pathological hallmarks of CBD. The injection of strain-specific tau aggregates does not only induce local tau disease but also spreads to distant brain regions known to be affected in CBD. These experiments suggest that there is intercellular transfer of strain-specific tau species in CBD which represent a putative target for therapy. There are several clinical trials with antibodies targeting tau. Gene therapy-based approaches to lower tau-burden are also being developed. The difficulties in proper diagnosis combined with the lack of robust biomarkers and a clinical rating scale have excluded CBD patients from the current tau-based trials (Svenningsson, 2019).

The baseline concentrations of NfL, in CSF or plasma, and CSF pTau are useful predictors of rates of change in clinical and neuropsychological status in PSP. These biomarkers were also associated with disease severity at baseline as reflected by clinical rating scales and neuropsychological measures of executive function scores. Similar to studies in other diseases and a previous study of plasma NfL, in PSP, higher CSF NfL concentrations were associated with greater disease severity and faster rates of disease progression. These data suggest that both CSF NfL and pTau are useful biomarkers to measure disease severity and predict change in PSP (Rojas et al., 2018).

TDP-43

Hyperphosphorylated transactive response DNA-binding protein 43 (TDP-43) is a proteinopathy associated with AD pathology. This protein increases with age and is related to an increased likelihood of expressing clinical Alzheimer's-type dementia. In the Rush Memory and Aging Project and Religious Orders Study of 946 older adults; 398 with and 548 without dementia, TDP-43 proteinopathy (cytoplasmic inclusions) was present in 496 (52%) subjects, and the pattern of deposition was classified as stage 0 (none; 48%), stage 1 (amygdala; 18%), stage 2 (extension to hippocampus/entorhinal; 21%), or stage 3 (extension to neocortex; 14%). TDP-43 pathology combined with a pathologic diagnosis of AD was a common mixed pathology (37% of all participants), and the proportion of subjects with clinical

Alzheimer's-type dementia formerly labeled "pure pathologic diagnosis of Alzheimer's disease" was halved when TDP-43 was considered. In logistic regression models adjusted for age, sex, and education, TDP-43 pathology was associated with clinical Alzheimer's-type dementia independent of pathological AD or other pathologies (infarcts, arteriolosclerosis, Lewy bodies, and hippocampal sclerosis). Mixed AD and TDP-43 pathologies were associated with higher odds of clinical Alzheimer's-type dementia than pathologic AD alone. In a large sample from almost 1000 community participants, TDP-43 proteinopathy was very common, frequently mixed with pathological AD, and associated with a higher likelihood of the clinical expression of clinical Alzheimer's-type dementia but only when extended beyond the amygdala (James et al., 2016; Nelson et al., 2019).

Normal pressure hydrocephalus

NPH is a disease of the elderly of unknown etiology (Adams et al., 1965). In the early reports, patients with NPH that underwent CSF shunting showed improvement in cognition, generating excitement that a "cure for dementia" had been found. However, the initial enthusiasm was dampened as many of the people that had shunts placed failed to improve and the incidence of complications was found to be high. The enlargement of the ventricles in many of these patients was due to loss of tissue based on other diseases, such as AD. Many of the patients diagnosed with NPH had hypertensive vascular disease, which could explain the lack of a response to shunting (Earnest et al., 1974; Graff-Radford & Godersky, 1987; Koto et al., 1977). NPH is often confused with vascular dementia because of the changes in the WM and the link with hypertension; this can make separation of the two diseases difficult.

NPH is a form of idiopathic communicating adult-onset hydrocephalus. Typically, patients with NPH have the triad of cognitive impairment, gait disturbance, and incontinence; similar symptoms are present in small vessel disease with extensive WM changes (Ojemann et al., 1969). There is slowing of the drainage of CSF with movement of water into the periventricular tissue, which is called transependymal absorption. The normal drainage routes of CSF fail to remove abnormal proteins and fluids because they are clogged secondary to infection or blood from subarachnoid hemorrhage. However, in about one-third of patients no etiology is found. Neuroimaging is used to identify those suspected to have NPH. Both CT and MRI are excellent ways to visualize the enlarged ventricles. MRI has the added advantage that the FLAIR sequences can show transependymal CSF absorption in a pattern of fluid surrounding the ventricles. As opposed to childhood forms of hydrocephalus, which are obstructive often due to aqueductal stenosis and accompanied by increased intracranial pressure that can be life-threatening, adult-onset hydrocephalus has normal or minimally elevated CSF pressure as measured during lumbar puncture. However, normal pressure is only partially correct; when patients undergo long-term monitoring, they are found to have intermittently elevated pressures, often during the night.

The presenting symptoms may be related to imbalance, incontinence, or impaired mental function. When gait is the presenting factor, the prognosis for treatment is better. NPH causes an apraxic gait, which is an inability to lift the legs, as if they were stuck to the floor. Motor strength is intact, reflexes are usually normal or slightly increased, and Babinski signs are absent; some patients have a grasp response of the toes when eliciting a Babinski sign, suggesting frontal lobe damage.

Another confounding diagnosis is small vessel WM disease because many of these patients also have hypertension with strokes, particularly lacunar state. Such patients may have other neurological findings including spasticity and hyperreflexia, while Babinski signs are rare in NPH, but present in small strokes. NPH leads to a reduction in intellect, which at times may be subtle. Dementia is of the subcortical type involves slowing of verbal and motor responses, with preservation of cortical functions such as language and spatial resolution. Neuropsychological testing quantifies the decline in intellect and the degree of dementia. Patients are apathetic and appear depressed. Incontinence of urine may occur, particularly in patients with prominent gait disturbance.

Diagnosis of adult-onset hydrocephalus and selection of patients for placement of a ventriculoperitoneal shunt has been difficult. Many of these patients have hypertensive vascular disease with lacunar infarcts. Features of Parkinson's disease were noted in earlier reports of the syndrome, and it is now recommended that all patients with Parkinson's disease have scans to rule out hydrocephalus. CT and MRI have aided in separating Parkinson's disease, lacunar state, and NPH, although NPH may occasionally coexist with these diseases. Patients diagnosed with vascular diseases, such as lacunar state or subcortical ischemic vascular disease with arteriosclerotic encephalopathy (Binswanger's disease) along with hydrocephalus, respond poorly to shunting, and if there is a positive response, it may be transient as the underlying disease progresses (Tullberg et al., 2002). Lumbar puncture (LP) with 20–40 mL of CSF removed often improves the gait, leading some investigators to consider the removal of CSF as a diagnostic test for placement of a shunt. Long-term drainage of the CSF by placing a lumbar catheter for continuous drainage improves with observation; to determine improvement in gait improves diagnostic accuracy. Finally, although not used routinely, radionuclide cisternography may be helpful; the radioligand injected into the CSF space in the lumbar region flows slowly over the convexity and enters the ventricles where it can remain trapped for several days. The selection of patients for shunting requires a combination of clinical findings and diagnostic test results; no test can predict whether a patient should undergo an operation. Drainage of CSF may involve the lymphatics of the brain, which are also called the glymphatics. Intrathecal injection of gadobutrol, a gadolinium contrast agent that can be imaged with T1-weighted MRI. There was reduced clearance of the contrast agent from the subarachnoid space and accumulation in the Sylvian fissure; this is similar to a radionuclide cisternography (Ringstad et al., 2017). A high incidence of vascular risk factors was found in patients with NPH, suggesting that it is a form of vascular dementia (Israelsson et al., 2017). Neuroimaging in patients with NPH has shown an enlargement of the temporal horns of the lateral ventricle with a disproportionate amount of cortical atrophy anticipated for the age of the patient. This is in contrast to patients with hydrocephalus ex vacuo due to a degenerative disease, such as AD, in which there is atrophy of the cerebral gyria and enlargement of both the sulci and ventricles. Another useful finding on FLAIR MRI is the presence of presumed transependymal fluid in the frontal and occipital periventricular regions.

As the number of patients who showed no improvement with shunts grew, and the complication rates of placing a shunt on an elderly patient became evident, the rate of diagnosis and number of shunts placed at most centers has dramatically declined. Unfortunately, none of the currently available tests by themselves identify the patient who will benefit from shunting. The most helpful is a combination of clinical signs and judiciously chosen laboratory tests. The decision to perform a shunt to treat NPH should be made cautiously since serious complications occurred in as many as one-fourth of the patients, including infection and subdural hematomas.

References

Adams, R. D., Fisher, C. M., Hakim, S., Ojemann, R. G., & Sweet, W. H. (1965). Symptomatic occult hydrocephalus with "normal" cerebrospinal-fluid pressure. A treatable syndrome. *The New England Journal of Medicine*, *273*, 117–126. Available from https://doi.org/10.1056/NEJM196507152730301.

Benveniste, H., & Nedergaard, M. (2021). Cerebral small vessel disease: A glymphopathy? *Current Opinion in Neurobiology*, *72*, 15–21. Available from https://doi.org/10.1016/j.conb.2021.07.006.

Boeve, B. F., Dickson, D. W., Duda, J. E., Ferman, T. J., Galasko, D. R., Galvin, J. E., Goldman, J. G., Growdon, J. H., Hurtig, H. I., Kaufer, D. I., Kantarci, K., Leverenz, J. B., Lippa, C. F., Lopez, O. L., McKeith, I. G., Singleton, A. B., Taylor, A., Tsuang, D., Weintraub, D., & Zabetian, C. P. (2016). Arguing against the proposed definition changes of PD. *Movement Disorders: Official Journal of the Movement Disorder Society*, *31*(11), 1619–1622. Available from https://doi.org/10.1002/mds.26721.

Caprihan, A., Raja, R., Hillmer, L. J., Erhardt, E. B., Prestopnik, J., Thompson, J., et al. (2021). A double-dichotomy clustering of dual pathology dementia patients. *Cerebral Circulation - Cognition and Behavior*, *2*, 1–15, 100011. Available from https://doi.org/10.1016/j.cccb.2021.100011.

Earnest, M. P., Fahn, S., Karp, J. H., & Rowland, L. P. (1974). Normal pressure hydrocephalus and hypertensive cerebrovascular disease. *Archives of Neurology*, *31*(4), 262–266.

Graff-Radford, N. R., & Godersky, J. C. (1987). Idiopathic normal pressure hydrocephalus and systemic hypertension. *Neurology, 37*(5), 868–871.

Hachinski, V. C., Iliff, L. D., Zilhka, E., Du Boulay, G. H., McAllister, V. L., Marshall, J., Russell, R. W., & Symon, L. (1975). Cerebral blood flow in dementia. *Archives of Neurology, 32*(9), 632–637.

Iba, M., Kim, C., Sallin, M., Kwon, S., Verma, A., Overk, C., Rissman, R. A., Sen, R., Sen, J. M., & Masliah, E. (2020). Neuroinflammation is associated with infiltration of T cells in Lewy body disease and α-synuclein transgenic models. *Journal of Neuroinflammation, 17* (1), 214. Available from https://doi.org/10.1186/s12974-020-01888-0.

Israelsson, H., Carlberg, B., Wikkelsö, C., Laurell, K., Kahlon, B., Leijon, G., Eklund, A., & Malm, J. (2017). Vascular risk factors in INPH: A prospective case-control study (the INPH-CRasH study). *Neurology, 88*(6), 577–585. Available from https://doi.org/10.1212/WNL.0000000000003583.

Jack, C. R., Bennett, D. A., Blennow, K., Carrillo, M. C., Dunn, B., Haeberlein, S. B., Holtzman, D. M., Jagust, W., Jessen, F., Karlawish, J., Liu, E., Molinuevo, J. L., Montine, T., Phelps, C., Rankin, K. P., Rowe, C. C., Scheltens, P., Siemers, E., Snyder, H. M., & Sperling, R. (2018). NIA-AA Research Framework: Toward a biological definition of Alzheimer's disease. *Alzheimer's & Dementia: The Journal of the Alzheimer's Association, 14*(4), 535–562. Available from https://doi.org/10.1016/j.jalz.2018.02.018.

James, B. D., Wilson, R. S., Boyle, P. A., Trojanowski, J. Q., Bennett, D. A., & Schneider, J. A. (2016). TDP-43 stage, mixed pathologies, and clinical Alzheimer's-type dementia. *Brain: A Journal of Neurology, 139*(11), 2983–2993. Available from https://doi.org/10.1093/brain/aww224.

Jellinger, K. A., & Korczyn, A. D. (2018). Are dementia with Lewy bodies and Parkinson's disease dementia the same disease? *BMC Medicine, 16*(1), 34. Available from https://doi.org/10.1186/s12916-018-1016-8.

Karanth, S., Nelson, P. T., Katsumata, Y., Kryscio, R. J., Schmitt, F. A., Fardo, D. W., Cykowski, M. D., Jicha, G. A., Van Eldik, L. J., & Abner, E. L. (2020). Prevalence and clinical phenotype of quadruple misfolded proteins in older adults. *JAMA Neurology*. Available from https://doi.org/10.1001/jamaneurol.2020.1741.

Karikari, T. K., Pascoal, T. A., Ashton, N. J., Janelidze, S., Benedet, A. L., Rodriguez, J. L., Chamoun, M., Savard, M., Kang, M. S., Therriault, J., Schöll, M., Massarweh, G., Soucy, J.-P., Höglund, K., Brinkmalm, G., Mattsson, N., Palmqvist, S., Gauthier, S., Stomrud, E., & Blennow, K. (2020). Blood phosphorylated tau 181 as a biomarker for Alzheimer's disease: A diagnostic performance and prediction modelling study using data from four prospective cohorts. *The Lancet. Neurology, 19*(5), 422–433. Available from https://doi.org/10.1016/S1474-4422(20)30071-5.

Kellar, D., & Craft, S. (2020). Brain insulin resistance in Alzheimer's disease and related disorders: Mechanisms and therapeutic approaches. *The Lancet. Neurology, 19*(9), 758–766. Available from https://doi.org/10.1016/S1474-4422(20)30231-3.

Koto, A., Rosenberg, G., Zingesser, L. H., Horoupian, D., & Katzman, R. (1977). Syndrome of normal pressure hydrocephalus: Possible relation to hypertensive and arteriosclerotic vasculopathy. *Journal of Neurology, Neurosurgery, and Psychiatry, 40*(1), 73–79.

Kovacs, G. G. (2016). Molecular pathological classification of neurodegenerative diseases: Turning towards precision medicine. *International Journal of Molecular Sciences, 17*(2). Available from https://doi.org/10.3390/ijms17020189.

Long, J. M., & Holtzman, D. M. (2019). Alzheimer disease: An update on pathobiology and treatment strategies. *Cell, 179*(2), 312–339. Available from https://doi.org/10.1016/j.cell.2019.09.001.

Maillard, P., Seshadri, S., Beiser, A., Himali, J. J., Au, R., Fletcher, E., Carmichael, O., Wolf, P. A., & DeCarli, C. (2012). Effects of systolic blood pressure on white-matter integrity in young adults in the Framingham Heart Study: A cross-sectional study. *The Lancet. Neurology*, *11*(12), 1039–1047. Available from https://doi.org/10.1016/S1474-4422(12)70241-7.

McKeith, I. G., Boeve, B. F., Dickson, D. W., Halliday, G., Taylor, J.-P., Weintraub, D., Aarsland, D., Galvin, J., Attems, J., Ballard, C. G., Bayston, A., Beach, T. G., Blanc, F., Bohnen, N., Bonanni, L., Bras, J., Brundin, P., Burn, D., Chen-Plotkin, A., & Kosaka, K. (2017). Diagnosis and management of dementia with Lewy bodies: Fourth consensus report of the DLB Consortium. *Neurology*, *89*(1), 88–100. Available from https://doi.org/10.1212/WNL.0000000000004058.

McKhann, G. M., Knopman, D. S., Chertkow, H., Hyman, B. T., Jack, C. R., Kawas, C. H., Klunk, W. E., Koroshetz, W. J., Manly, J. J., Mayeux, R., Mohs, R. C., Morris, J. C., Rossor, M. N., Scheltens, P., Carrillo, M. C., Thies, B., Weintraub, S., & Phelps, C. H. (2011). The diagnosis of dementia due to Alzheimer's disease: Recommendations from the National Institute on Aging-Alzheimer's Association workgroups on diagnostic guidelines for Alzheimer's disease. *Alzheimer's & Dementia: The Journal of the Alzheimer's Association*, *7*(3), 263–269. Available from https://doi.org/10.1016/j.jalz.2011.03.005.

McKhann, G., Drachman, D., Folstein, M., Katzman, R., Price, D., & Stadlan, E. M. (1984). Clinical diagnosis of Alzheimer's disease: Report of the NINCDS-ADRDA Work Group under the auspices of Department of Health and Human Services Task Force on Alzheimer's Disease. *Neurology*, *34*(7), 939–944.

Nelson, P. T., Dickson, D. W., Trojanowski, J. Q., Jack, C. R., Boyle, P. A., Arfanakis, K., Rademakers, R., Alafuzoff, I., Attems, J., Brayne, C., Coyle-Gilchrist, I. T. S., Chui, H. C., Fardo, D. W., Flanagan, M. E., Halliday, G., Hokkanen, S. R. K., Hunter, S., Jicha, G. A., Katsumata, Y., & Schneider, J. A. (2019). Limbic-predominant age-related TDP-43 encephalopathy (LATE): Consensus working group report. *Brain: A Journal of Neurology*, *142*(6), 1503–1527. Available from https://doi.org/10.1093/brain/awz099.

Ojemann, R. G., Fisher, C. M., Adams, R. D., Sweet, W. H., & New, P. F. (1969). Further experience with the syndrome of "normal" pressure hydrocephalus. *Journal of Neurosurgery*, *31*(3), 279–294.

Panza, F., Lozupone, M., Seripa, D., Daniele, A., Watling, M., Giannelli, G., & Imbimbo, B. P. (2020). Development of disease-modifying drugs for frontotemporal dementia spectrum disorders. *Nature Reviews. Neurology*, *16*(4), 213–228. Available from https://doi.org/10.1038/s41582-020-0330-x.

Ringstad, G., Vatnehol, S. A. S., & Eide, P. K. (2017). Glymphatic MRI in idiopathic normal pressure hydrocephalus. *Brain*, *140*(10).

Rojas, J. C., Bang, J., Lobach, I. V., Tsai, R. M., Rabinovici, G. D., Miller, B. L., & Boxer, A. L. (2018). CSF neurofilament light chain and phosphorylated tau 181 predict disease progression in PSP. *Neurology*, *90*(4), e273–e281. Available from https://doi.org/10.1212/WNL.0000000000004859.

Rosenberg, G. A., Prestopnik, J., Knoefel, J., Adair, J. C., Thompson, J., Raja, R., & Caprihan, A. (2019). A Multimodal approach to stratification of patients with dementia: Selection of mixed dementia patients prior to autopsy. *Brain Sciences*, *9*(8). Available from https://doi.org/10.3390/brainsci9080187.

Rosenberg, G. A., Wallin, A., Wardlaw, J. M., Markus, H. S., Montaner, J., Wolfson, L., Iadecola, C., Zlokovic, B. V., Joutel, A., Dichgans, M., Duering, M., Schmidt, R., Korczyn, A. D., Grinberg, L. T., Chui, H. C., & Hachinski, V. (2016). Consensus statement for diagnosis of subcortical small vessel disease. *Journal of Cerebral Blood Flow and Metabolism: Official Journal of the International Society of Cerebral Blood Flow and Metabolism, 36*(1), 6–25.

Snowdon, D. A. (2003). Healthy aging and dementia: Findings from the Nun Study. *Annals of Internal Medicine, 139*(5 Pt 2), 450–454.

Svenningsson, P. (2019). Corticobasal degeneration: Advances in clinicopathology and biomarkers. *Current Opinion in Neurology, 32*(4), 597–603. Available from https://doi.org/10.1097/WCO.0000000000000707.

Thomas, C. L. (Ed.), (1985). *Taber's Cyclopedic Medical Dicdtionary* (15th). Philadelphia: F.A. Davis Company.

Toledo, J. B., Arnold, S. E., Raible, K., Brettschneider, J., Xie, S. X., Grossman, M., Monsell, S. E., Kukull, W. A., & Trojanowski, J. Q. (2013). Contribution of cerebrovascular disease in autopsy confirmed neurodegenerative disease cases in the National Alzheimer's Coordinating Centre. *Brain: A Journal of Neurology, 136*(Pt 9), 2697–2706. Available from https://doi.org/10.1093/brain/awt188.

Tomimoto, H. (2011). Subcortical vascular dementia. *Neuroscience Research, 71*(3), 193–199. Available from https://doi.org/10.1016/j.neures.2011.07.1820.

Tomlinson, B. E., Blessed, G., & Roth, M. (1970). Observations on the brains of demented old people. *Journal of the Neurological Sciences, 11*(3), 205–242.

Tullberg, M., Hultin, L., Ekholm, S., Månsson, J. E., Fredman, P., & Wikkelsø, C. (2002). White matter changes in normal pressure hydrocephalus and Binswanger disease: Specificity, predictive value and correlations to axonal degeneration and demyelination. *Acta Neurologica Scandinavica, 105*(6), 417–426. Available from https://doi.org/10.1034/j.1600-0404.2002.01189.x.

Yang, M., & Williamson, J. (2019). Blood pressure and statin effects on cognition: A review. *Current Hypertension Reports, 21*(9), 70. Available from https://doi.org/10.1007/s11906-019-0973-4.

Chapter 2

Neuroimaging methods

Historical perspective

Little progress was made in understanding dementias of old age since the end of the 19th century, when Alois Alzheimer described the plaques and tangles in the brain of a patient with early onset of severe dementia, until the 1970s. During that long time interval, Alzheimer's disease (AD) was classified as presenile dementia in those less than 65 years of age and senile dementia in those over 65. The term "senile," which is no longer used, had the implication that it was a disease of old age and that everyone as they aged was at risk of developing it. As awareness grew that memory loss was not an inevitable occurrence of aging, the use of the term "senile" was abandoned. We now refer to early onset AD, which generally has a genetic basis and late onset AD, which occurs sporadically.

Our understanding of AD began to evolve during the 1970s, as advances in molecular biology and neuroimaging greatly improved our ability to diagnose dementia during life and subsequently to confirm it pathologically. This ability to diagnose AD during life has led to a marked increase in the number of patients and a greater understanding of multiple etiologies. There have been major advances in the use of biomarkers and improvements in neuroimaging that have expanded the number of diagnoses of dementia. Parallel with the increased understanding of AD has been an increased understanding of other causes of dementia, which has led to the use of the term, AD and related dementias. Molecular biology, genetics, and artificial intelligence have produced an explosion of information on the cognitive illnesses of aging, which is now beginning to move toward effective treatments.

The most important advances that began in the 1970s were made in neuroimaging. The first was the invention of computed tomography (CT) in England by Jeffrey Housefield; he used X-rays to construct tomographic slices of the brain and revolutionized clinical medicine. The first computer tomographic instruments showed brain pathology only available at autopsy prior to that time. It revealed brain atrophy, enlarged ventricles, ischemic injury, brain tumors, and damage to white matter that were not seen prior to CT. From the early days of neuroimaging, an amazing number of applications has rapidly emerged by increasing the number of pixels, making structures more clearly visible. Housefield's discovery was applied to other

Neuroinflammation in Vascular Dementia. DOI: https://doi.org/10.1016/B978-0-12-823455-6.00008-0

imaging methods, including nuclear magnetic resonance (NMR) imaging (MRI) and isotopic studies with positron emission tomography (PET). These imaging devices have greatly increased understanding of brain structure and function.

Shortly after the discovery of CT and its routine use in clinical practice, an article was published where a patient had extensive attenuation (loss) of white matter signal on the CT. This patient had gait problems, incontinence, apathy, and early dementia. At autopsy, he was found to have demyelination with hypertensive damage to the arterioles, leading to a diagnosis of Binswanger's disease (BD). This was the first patient to have the disease of the white matter diagnosed by CT, and subsequently shown at autopsy. The CT showed extensive damage in frontal, occipital, and semiovale white matter regions (Fig. 2.1) (Rosenberg et al., 1979). With the increased use of CT in diagnosis of dementia, there was a growing number of people with normal cognition that were diagnosed with BD on the basis of the CT. In 1981 an editorial in *Lancet*, the British medical journal, said there was an epidemic of BD based on the CT (Binswanger's Encephalopathy, 1981). Basing a diagnosis on a radiographic test was discouraged since patients were distressed to learn that they had a dementing illness. A study done on healthy normal older individuals over the age of 65 confirmed the high incidence of white matter changes in an MRI; in 100 normal subjects over the age of 65, 30% had moderately severe white matter lesions on fluid attenuated inversion recovery (FLAIR) MRI (Hunt et al., 1989).

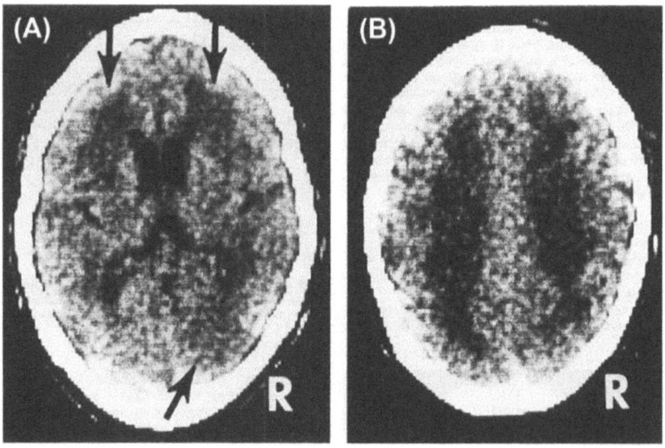

FIGURE 2.1 **CT of Binswanger's disease.** CT scan from first admission. (A) Sharply marginated symmetrical areas of abnormal low density in the white matter of the frontal lobes (*vertical arrows*) and a more vague low-density abnormality in the right occipital area (*oblique arrow*). (B) A higher cut superior to the lateral ventricles. Note extensive low-density abnormality throughout the centrum semiovale of the two hemispheres. Enhancement was not seen after intravenous contrast medium administration. *From Rosenberg, G. A., Kornfeld, M., Stovring, J. & Bicknell, J. M. (1979). Subcortical arteriosclerotic encephalopathy (Binswanger): computerized tomography.* Neurology, 29*(8), 1102–1106.*

The application of tomography to MRI revealed in exquisite detail atrophy of the cortex and hippocampus in patients with dementia. But more importantly, it showed that damage to the white matter structure was a major factor in dementia, particularly vascular and inflammatory causes of dementia; this resulted in dementia evolving from mainly a focus on diseases involving the gray matter to an increased awareness of the importance of the white matter in dementia. Within a short time after the introduction of MRI, vascular disease of white matter, first described pathologically by Otto Binswanger, was recognized as a common disease of the elderly that had been largely ignored by the lack of interest in autopsies in elderly patients.

Routine use of CT and MRI in neurological practice beginning in the late 1970s made diagnosing dementias possible prior to autopsy, and greatly enhanced clinical diagnosis by combining clinical evaluation with neuroimaging. When patients rarely lived beyond 50 or 60 years of age and when few autopsies were done, the diseases described by Alzheimer and Binswanger were considered rare, and patients with cognitive impairment were considered to have "hardening of the arteries," which we now call vascular dementia. As the number of patients thought to have Alzheimer type dementia increased, the need for more standardized diagnostic criteria was evident (McKhann et al., 1984). Early criteria were proposed without neuroimaging, which was added later to aid classification of patients with dementia (Chui et al., 1992).

While AD eventually involves all aspects of cognition, the most common early symptoms relate to memory, spatial orientation, and language. Diagnosis of patients with AD has been aided by a progressive course with deterioration primarily in memory, which can be repeatedly measured with tests, such as the mini-mental status examination (MMSE) and the Montreal Cognitive Assessment (MoCA). The overlapping of symptoms in patients with dementia confounded the ability to reach a clinical diagnosis. To overcome this problem, the use of amyloid and tau biomarkers derived from CSF and PET studies has been proposed (Jack et al., 2018).

As opposed to the relentless course of AD, the course of vascular causes of dementia is slower with long periods of remission. Initially, it was considered due to strokes and it was thought that a large volume of brain tissue had to be injured in order to lead to dementia. This concept persisted for some time, since autopsies were uncommon in older individuals, and death was attributed to consequences of atherosclerosis. After CT and MRI came into routine use, the concept of small strokes rather than large strokes became more prevalent since the consequences of small vessel disease (SVD) could be seen on CT and MRI. The fact that damage to the white matter was commonly seen on MRI raised interest in vascular dementia.

As vascular diseases are increasingly recognized as important in dementia, a number of terms have been introduced to define vascular causes of dementia. Multi-infarct dementia (MID) was initially the preferred term

(Hachinski et al., 1975). However, this proved to be inadequate when many patients lacked evidence of strokes on MRI, but rather were observed to have injury to the deep white matter. The number of pathological processes that were associated with cognitive loss resulted in a confusing array of diagnostic terms. Researchers could not agree on what to call this multiple syndrome process. A breakthrough came in 2006 with a conference organized by researchers in Canada and the United States to reach a consensus, or a "harmonization," as to what terms should be used to define vascular aspects of dementia (Hachinski et al., 2006). Rather than attempt to put all forms of pathology into one diagnostic category, it was realized that multiple pathological processes could lead to the same result. The term that emerged from that consensus conference was "vascular cognitive impairment (VCI)." This included small vascular ischemic events, referred to as lacunar strokes, large vessel strokes, and a new concept of subcortical ischemic vascular disease (SIVD) to indicate extensive damage to the deep white matter. With this new formulation, the variety of underlying pathology could now be studied. This had a major impact on the field because the arguments over nomenclature, although by no means entirely resolved, became less important than an attempt through neuroimaging and pathology to determine more precisely the underlying disease process. Simultaneously, there was a growing awareness of the importance of white matter damage not necessarily associated with discreet strokes. This SVD was the result of vascular risk factors that had the greatest impact on the arterioles. The MRI appearance of SVD was similar to that seen in multiple sclerosis (MS). The MRI pulse sequence that showed the increase in water in the tissue was Fluid attenuated inversion recovery (FLAIR). White matter hyperintensities (WMHs) were seen routinely in MS patients, but these nonspecific changes were present in a number of vascular and nonvascular neurological diseases. In addition, a high percentage of healthy people were also found to have these changes on MRI, making it difficult to use WMHs as a biomarker of brain disease, but those with WMHs were found to have a higher incidence of vascular risk factors. In spite of the increase with vascular factors, the overlap of pathological changes with normal people greatly reduced the importance of WMH findings on MRIs. In initial neuroimaging, the presence of attenuation of the white matter on CT or of WMHs on MRI was thought to be pathological. However, healthy elderly people had abnormal white matter according to CTs and this was accentuated with the use of the more sensitive MRI. When the changes in the white matter were large and confluent, a diagnosis of BD was made based on the earlier pathological findings of vascular disease-associated demyelination.

The problem of overdiagnosis created by MRI was partially resolved by two novel magnetic resonance methods that indicated tissue injury, adding a deeper meaning to the WMHs. The first method is proton NMR spectroscopy (^{1}H-MRS): essentially using magnetic resonance as a spectrometer that could quantitate individual molecules in the tissue. This was possible because the molecular spectrum of each molecule was distinct. When the tissue was injured or accumulated abnormal amounts of a substance, the ^{1}H-MRS spectrum was altered, revealing the extent of the injury and the amount of the major molecules that were involved. The brain contains a large amount of *N*-acetylaspartate (NAA), and when the tissue is damaged as occurs in hypoxia/ischemia, there is a decrease in the signal from NAA proportional to the fall in concentration. The second MR method that extended the usefulness of the MRI was diffusion tensor imaging (DTI); when the white matter was injured, the movement of water in the tissue was altered, and this change could be detected by MR methods. White matter has NAA in the axons, which is a common site of injury by hypoxia. Thus white matter injury could be detected by a fall in the NAA signal with ^{1}H-MRS and with a change in diffusion of water in the tissue as detected by DTIs.

Discovery of computerized axial tomography

Developments of neuroimaging were based on the fundamental discovery in the 1970s of methods to take tomographic images of objects: computerized axial tomography was an X-ray technique based on the ability to rotate the gantry and to solve over 100,000 simultaneous differential equations that determine the density of the x-rays at each location in the image, which allows a reconstruction of the pixels to form the image. These discoveries were made by engineers working at the British record company Electric and Musical Industries (EMI) and the original instruments were called EMI scanners. The inventors, Allan McLeod Cormack and Godfrey Newbold Hounsfield, were awarded the Nobel Prize in Physiology and Medicine in 1979. The CT utilized X-rays to create tomographic images that were initially of very low resolution, but were still able to distinguish between gray and white matter and to visualize the ventricles, since the X-rays penetrated different tissues at different rates. Bone stopped the X-rays, while water and air had relatively little effect. It is also able to show attenuation of the white matter. Early CT scans were able to visualize ventricular enlargement, making it excellent for hydrocephalus. Image quality improved as the resolution of CT scanners increased over several years. The original studies of dementia occurred with vascular dementia, since the injured white matter could be seen as an "attenuated" signal based on the increase in water or loss of myelin. This resulted in early descriptions of BD patients with vascular demyelination (Caplan & Schoene, 1978). The CT scanners were soon replaced by MRI scanners that utilized the same tomographic principals, speeding up the pace of technological advances and improving diagnostic accuracy.

The tomographic method was used with other modalities; radiolabeled isotopes were shortly afterwards used with positrons to form PET, which could identify specific molecules. Fluorodeoxyglucose (FDG) is a PET molecule of glucose that is trapped after initiating metabolism showing the regions that were undergoing glucose usage (Sokoloff, 1984). FDG scans in patients with AD showed hypometabolic regions. Besides showing metabolism, ligands of important brain molecules could be linked with the positrons and used to show the location and quantity of the molecules of interest. Synthesizing ligands for amyloid, tau, and more recently α-synuclein was a major breakthrough for studies of dementia. Recent clinical trials have used PET as a surrogate biomarker to show removal of amyloid and tau (Klunk et al., 2001). Because of the high cost of the instruments and the need for a nuclear chemistry laboratory to produce the ligands, PET studies were originally done at large research centers that had the instruments to record positron signals and the adjacent radiochemical laboratory that could produce the compounds. Today, PET studies are limited to a number of research centers. Currently, once the positron-labeled compounds are produced, they need to be used in a short time based on the half life. Since positrons have different half-lives, those with longer half-lives can be produced remotely and shipped to research centers, making PET scanning more available.

While the conceptual basis for NMR has been known for many years and used extensively for unraveling molecular structure and understanding chemical reactions, it was not until the 1980s that the concept of using MR to construct tomographic images of the brain actually entered into clinical practice. Peter Mansfield and Paul Lauterbur received the Nobel Prize in Physiology and Medicine in 2003 for their independent discovery of the use of MR for imaging. The greater resolving power and multiple uses in understanding brain physiology along with structure, made the MRI the most important of the new devices, replacing CT in developed countries.

Nuclear magnetic resonance spectroscopy and magnetic resonance imaging

Credit for the discovery of NMR goes to Isidor Isaac Rabi, who received the Nobel Prize in Physics in 1944. Rabi described how nuclei could be induced to flip their principal magnetic orientation by an oscillating magnetic field. In 1946, two scientists in the United States, independently of each other, described a physicochemical phenomenon that was based upon the magnetic properties of certain nuclei in the periodic system. The next major discovery in NMR was made by Felix Bloch and Edward Mills Purcell, who applied the technique for use on liquids and solids, for which they shared the Nobel Prize in Physics in 1952. Rabi, Bloch, and Purcell observed that magnetic nuclei, like ^{1}H and ^{31}P, could absorb radiofrequency (RF) energy when placed in a magnetic field, and when the frequency specific to the identity of the nuclei energy was absorbed, the nucleus was described as being *in resonance*. Different atomic nuclei within a molecule resonate at different (radio) frequencies for the same magnetic field strength. The observation of such magnetic resonance frequencies of the nuclei present in a molecule allows any trained user to discover essential chemical and structural information about the molecule. Several investigators led by Gadian and Rada in England recognized the usefulness of NMR for the study of biological systems; these original studies were done with ^{31}P in energy metabolism, utilizing the separate resonances of the three phosphorus molecules in adenotriphosphate (ATP) (Hoult et al., 1974).

Biological NMR began with the study of metabolism, which was possible by utilizing atoms with magnetic moments contained in molecules involved in metabolism. The earliest atoms used were ^{13}C, ^{19}F, ^{23}Na, and ^{1}H, since these atoms have magnetic moments that can be detected with an NMR spectrometer. When molecules with magnetic moments are incorporated into molecules, such as ^{13}C-glucose, it is possible to follow that molecule through an entire metabolic cycle, as the ^{13}C-glucose moves from one molecule to another and the resonance patterns change with each location of ^{13}C in the molecule (Brainard et al., 1989). Since magnetic fields are innocuous, the use of MRS provided a means to noninvasively measure brain metabolism repeatedly and to map metabolism over time. However, the major limitation to the more general use of NMR for biochemical studies in vivo was the inherent weakness of the signal, making detection of the molecules difficult. Incorporating atoms with magnetic moments into molecules and the long times necessary to record the signals made the use of NMR in studies of biochemical processes expensive and time-consuming. Although the study of metabolism is important, the most exciting use of the methods of computed tomography in MRI was its application in the construction of MRI, which has made the greatest impact because of the abundance of ^{1}H in water molecules, creating strong NMR signals and allowing for the exquisite anatomical detail in MRI images. Furthermore, by modifying the pulse sequences, MR methods were expanded to show physiological function including cerebral blood flow, molecular composition with magnetic resonance spectroscopy, and dynamic brain function with resting state functional MRI (fMRI). Rapid advances in computer applications have made visualization of individual fiber tracts possible, leading to the ability to reconstruct the "connectome"; revealing the wiring of the brain. With appropriate pulse sequences, studies of diffusion of water in the brain made the mapping of the fiber tracts possible; when diffusion is restricted by cellular swelling and other forms of injury, there is a change in the diffusion signal that occurs at the earliest stages of the injury. Water movement as measured by diffusion pulse sequences has a directional component that is more evident in the white matter because of the straight pathways between white matter fiber tracts. When there is a directional component to the diffusion signal, diffusion is called anisotropy. On the other hand, nondirectional, random movement of water is called isotropic diffusion. The latter is more common in the dense neuropil of the gray matter. Taking advantage of the directionality imposed by fiber tracts, it is possible to reconstruct this type of water movement to delineate fiber bundles. This is called tractography. More importantly for clinical studies, white matter injury produces abnormal signals in the MRI, particularly in the routine clinical sequence called FLAIR, which is a pulse sequence that suppresses water in the ventricles and accentuates the water in the tissue allowing visualization of water in tissue.

Basic concepts in nuclear magnetic resonance

Understanding the physics behind the biological use of NMR is beyond the scope of this book since it requires a knowledge of quantum mechanics. However, the basic concepts can be grasped without resorting to mathematical formulations. Therefore what follows is a simplified discussion focused on those basic aspects necessary for an understanding of MRI.

When nuclei with magnetic moments are placed in a magnetic field, they align themselves parallel to the field. Atoms have nuclear spins whose value depends on the mass number and the atomic number of the nucleus. Those nuclei with spins of 1/2, 3/2, 5/2, and so on will resonate when placed in an applied magnetic field; the resonant frequency will depend on the magnetogyric ratio, γ, which is a proportionality constant specific for each particular atom. The fundamental resonance condition for all NMR experiments is given by the Larmor equation: $n_o = \gamma B_o/2p$ where n is the frequency (Hz) at which a nucleus of a given magnetogyric ratio resonates when a magnetic field, B_o, is applied. The field or the frequency may be varied in order to obtain spectra. The strength of a magnetic field is given in Tesla (T) or kilogauss (kG) (1 T = 10 kG). When nuclei with magnetic moments are exposed to radio waves of a specific frequency, they emit a signal of the same frequency as that applied. Atoms with magnetic moments can assume one of several discrete energy states when placed in a magnetic field. When another magnetic field is applied, the atom undergoes a transition from one energy state to another. Energy levels in NMR are in the radio frequency range of 107 Hz. Energy is absorbed and emitted in discrete packets. Increasing the energy state causes a transition to a higher level. Because the magnetic forces of atoms are small, the amount of energy needed to raise the energy state is in the 105–107 Hz range. X-rays, in comparison, excite electrons at 1017 Hz. The strength of the magnetic field can vary over a wide range, from very low fields of 0.05 T to high fields of 11 T.

Superconducting helium-cooled magnets are needed to produce a higher field strength for MRI; the higher the field, the better the resolution of the image and the spectra. An added advantage of high-field magnets is that, as the field strength increases, the signal-to-noise ratio improves, which is particularly important in NMR spectroscopy, providing sharper peaks with better spectral definition. When the NMR instrument is used to perform spectroscopic analysis, other nuclear in addition to protons, such as ^{31}P,

^{23}Na, ^{13}C, and ^{15}N, can be detected; molecules containing these atoms have weaker signals than those seen with water protons because of their much lower abundance. Electrons in orbits around the atoms modify the magnetic signal from the atom; this is called a chemical shift. Since hydrogen has only one orbital electron, it has the greatest response to the applied magnetic field. That is, the proton experiences a greater effect than an atom with more electrons and, therefore more atomic shielding (Fig. 2.2).

For example, hydrogen with a spin of 1/2 has two possible positions $(2j + 1)$, where j is the spin number. An NMR spectrometer works by inducing an oscillating field around the sample at a frequency that causes nuclei in the sample to resonate. When the resonating frequency induced equals the characteristic frequency for that substance, there is a transition from the lower to the higher energy state and energy is absorbed. Normally, there is a small preponderance of nuclei that are in the lower energy state. This amounts to less than 1 part in 10^6. The slight energy absorption can be detected with a sensitive electronic amplifier. The resonance frequency causes the nuclear magnets to precess synchronously. The spectrometer detects the signals, which begin to decay with time. Fourier transforms convert the signals from the time domain to the frequency domain, allowing for the simultaneous recording of multiple frequencies in the same magnetic field. By turning the signal on and off rapidly at a time interval of t seconds, a range of frequencies centered on the original frequency, F, occurs. When the pulse is applied, a single decaying curve is obtained in the time domain, which converts to a single peak in the frequency domain. Relaxation of molecules when the pulse is stopped leads to the return to the unexcited state, that is, the alignment of the nuclei in the field is lost. Relaxation to the unexcited state is referred to as *T1 relaxation*. Interactions between nuclei also occur, and relaxation of the interacting nuclei is referred to as *T2*

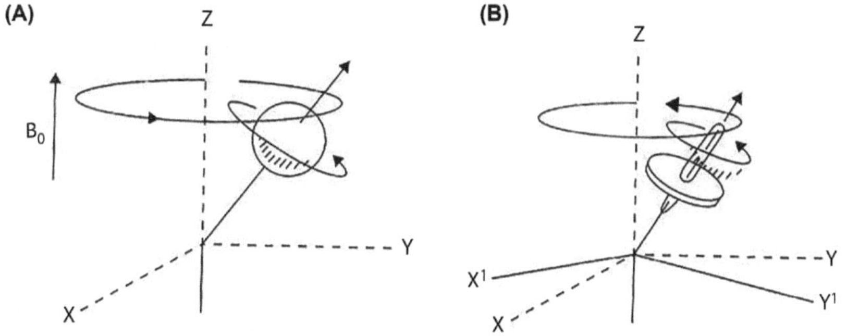

FIGURE 2.2 (A) Protons of hydrogen have a spin of 1/2 and two possible energy states. As they line up in the field, they experience a force that causes them to gyrate about their axis. (B) The angular velocity at which they gyrate or precess depends on the field strength and the characteristics of the nuclei.

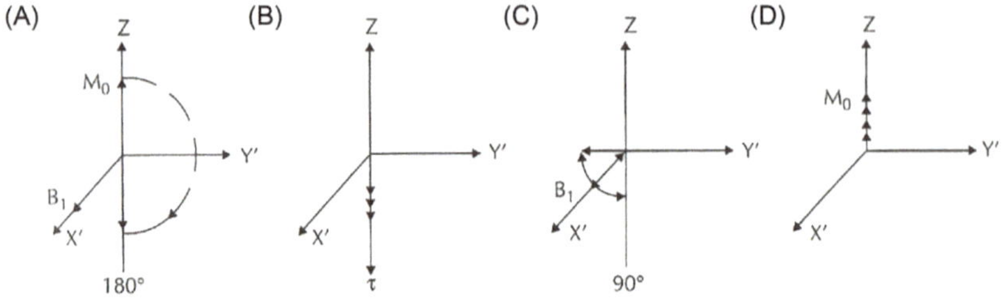

FIGURE 2.3 Inversion-recovery Fourier transform method for T1 determination. (A) Net magnetization in the z direction equals M_0, a 180-degree pulse inverts the magnetization vector. (B) M_z then recovers toward its initial value, M_0, during time τ. (C) A 90-degree pulse tilts this magnetization into the recording *xy*-plane. The size of the resulting signal is determined by the magnetization at the time of recording. (D) After a delay of time $D > 4T1$, the magnetization, M_z, has relaxed to its initial value of M_0 and the cycle can be repeated. *From Gadian, D. G. (1982). Nuclear magnetic resonance and its application to living systems. Oxford: Oxford University Press.*

relaxation (Fig. 2.3). T1 and T2 are used extensively in image formation. A T1-weighted MRI scan has a shorter relaxation time and shows signals in protons attached to molecules in membranes, while the longer-relaxing molecules that remain spinning are not observed. Thus in the T1-weighted image, water appears as a dark region and the tissue is lighter, which can be readily visualized by the very dark regions inside the water-filled cerebral ventricles. By contrast, T2-weighted images are formed at longer relaxation times, when the water molecules are still spinning, resulting in detection of the molecules with long resonances. The technique of allowing the spinning atoms to return to the original state is referred to as "spin-echo" because it refocuses the water molecules during the longer recording times (Fig. 2.4).

Protons on the inelastic membranes lose resonance, like a spring returning to their original position, while those in the water continue to spin, allowing them to be detected during the spin echo. An important addition to the imaging sequences was the use of FLAIR, which uses shorter imaging times to show water that is relaxing at an intermediate speed. This allows the water in the tissues to be seen, revealing edema fluid clearly, but reduces the signal from the water in the ventricles, which is still spinning. The FLAIR sequence is an important improvement that shows water in tissues while suppressing the water proton signal in CSF. This is an excellent sequence to detect the transependymal flow of water in hydrocephalus and early edema formation in the ischemic brain.

Spin-lattice or T1 recovery can be measured by the inversion-recovery method. The original alignment in the field along the B_o field lines results in alignment of the nuclei in the z direction. A pulse of RF energy of sufficient length to invert the magnetic moment 180 degrees is applied. After a waiting period of length τ, the signal begins to recover. Another pulse of 90 degrees

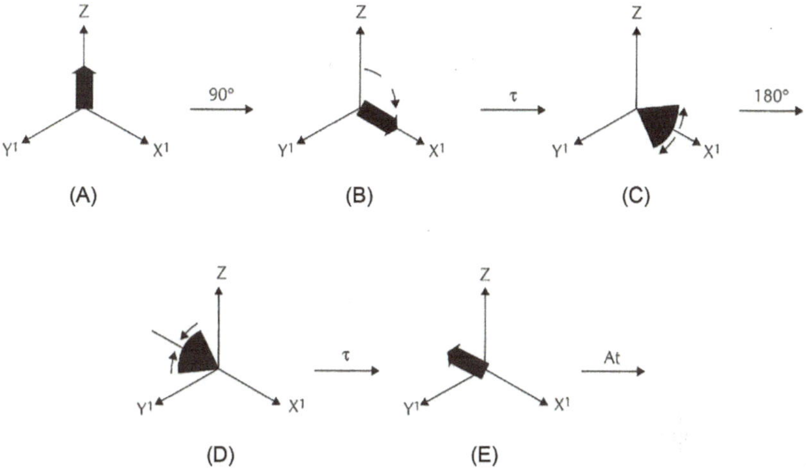

FIGURE 2.4 T2-weighted images are formed at longer relaxation times, when the water molecules are still spinning, resulting in detection of the molecules with long resonances. The technique of allowing the spinning atoms to return to their original state is referred to as spin-echo because it refocuses the water molecules during the longer recording times. Spin-echo T2 measurement. (A) Magnetization in the z-direction. (B) A 90-degree pulse tilts the magnetization onto the x'-axis. (C) The field inhomogeneity causes the vector to diphase or fan out over time. (D) A 180-degree pulse about the x'-axis reverses the magnetization. (E) The precessing atoms come together or refocus into an echo during time τ. *From Gadian, D. G. (1982). Nuclear magnetic resonance and its application to living systems. Oxford: Oxford University Press.*

is given, and the amount of signal that has recovered is recorded during the collection time, *tD*. Signal collection occurs in the *y*-plane; the 90 degrees pulse is needed after the 180 degrees inversion pulse to position the signal in the correct plane. An inversion-recovery pulse sequence is given as (180°−p−90°− *AT*−*D*)*n* where *AT* is the acquisition time for the signal to be collected and *D* is the delay between collections, which is at least four times the *T*1 value. *T*1 measurements are made by collecting spectra at different *t* values and plotting the value against *T*. A straight line results with a slope of 1/*T*1. The interaction between molecules is measured by the T2 or spin-spin relaxation methods. Spin−spin measurements are more difficult to do and more subject to error than T1. Measurements of T2 relaxation are done by the spin-echo method. The magnetization is tilted 90 degrees into the *xy*-plane. The phase coherence is gradually lost because of slight inhomogeneities in the field. This results in the faster-moving molecule dephasing ahead of the slower-moving one. When the field is flipped 180 degrees, the slower-moving nuclei are placed inside the faster-moving ones so that they refocus at the same time, partially correcting the effect of the inhomogeneity. After a time delay, the signal was dephased around the *z*-axis. Another pulse of 180 degrees reverses the dephasing signal, and continued fanning brings the signal to a focus along the *y*-axis, forming a mirror image or echo of the original signal. The echo can be repeated a number of times every 2*t* seconds. As with the T1 measurements, over time the signal intensity at each echo is slightly less and T2 can be calculated. Relaxation measurements are made after pulse sequences have been applied to help identify the origin of the signals. Bound molecules, such as those in crystals or membranes, have shorter relaxation times because they are less mobile than unbound ones. Therefore altering the delay time before signal acquisition after the final pulse in the sequence determines the components included in the signal. For example, a long delay time removes the broad signals from bound membrane compounds since they have relaxed early and allows the selective acquisition of the smaller, more mobile metabolites that take longer to fully relax.

MRI has gradients that can be applied along three axes: *x*, *y*, and *z*. These gradients separate the field into voxels and the signal in each voxel can be determined to form an image as done with computed tomography. The ability to localize signals to brain regions is the basis for the construction of DTI that is used to construct fiber tracts and to perform functional MRI in three directions.

In FLAIR sequences, an inversion-recovery pulse is used to null the signal from cerebrospinal fluid. When the net magnetization vector of the cerebrospinal fluid passes its null point, little or no longitudinal magnetization is present in the fluid. The transverse magnetization of cerebrospinal fluid is insignificant, and therefore no signal is generated. Elimination of the signal from cerebrospinal fluid is useful for detecting lesions that otherwise are not easily distinguishable or for delineating hyperintense lesions that border fluid-containing spaces such as sulci or ventricles in the brain (Bitar et al., n.d.).

FLAIR MRI shows WMHs in patients with vascular dementia, but is unable to distinguish between an increase in water from edema and a loss of myelin due to loss of myelinated fibers. A typical patient with BD, the small vessel form of vascular dementia, has extensive WMHs (Fig. 2.5A): the water-filled ventricles are dark and the white matter has increased signal. The T1-weighted image (Fig. 2.5B) has atrophy and evidence of lacunar infarcts in the basal ganglia,

The exquisite sensitivity of FLAIR images to demonstrate hyperintense regions of white matter creates problems in very old healthy individuals, since they can have increased white matter signals as part of the aging process. Because the white matter is susceptible to these changes, a term, leukoaraiosis, has been proposed to describe "pale" white matter; the original intent of the term was that it be used without attaching any pathological significance to the term unless symptoms consistent with dementia are present (Hachinski et al., 1987). MRI and neuropsychological testing were done in a study of 100 healthy elderly individuals over the age of 65. MRIs showed that 30% had moderately severe WMHs with normal scores on

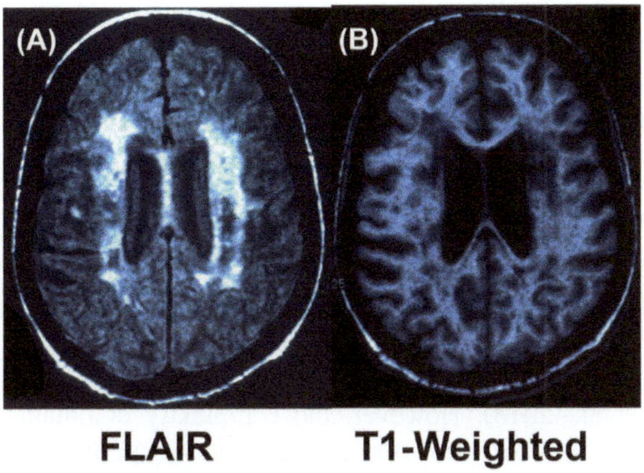

FIGURE 2.5 **MRI of patient with Binswanger's disease.** (A) FLAIR image of a patient with Binswanger's disease. The periventricular white matter is seen as white, indicating that it is injured. There is no ventricular enlargement nor cortical atrophy. (B) A T1-weighted image that shows the white matter injury as a dark area around the ventricle. Some tissue loss can be seen better on the T1 image than on the FLAIR. *FLAIR*, Fluid attenuated inversion recovery; *MRI*, magnetic resonance imaging.

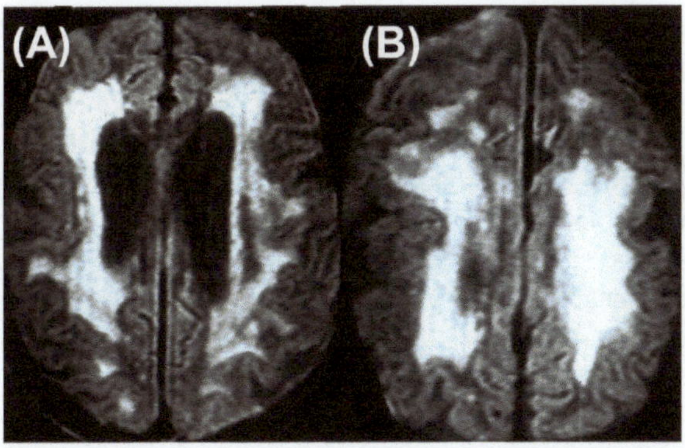

FIGURE 2.6 MRI of healthy individuals over 90 years of age with white matter changes of unknown etiology is referred to as leukoaraiosis. Almost all of the white matter is affected. (A) At the level of the lateral ventricles. (B) At a higher level, including the centrum semiovale, the extent of the white matter change can be clearly seen. *MRI*, Magnetic resonance imaging.

neuropsychological testing; seven of the oldest in the group had complete white matter signal increase on MRI (Fig. 2.6) (Hunt et al., 1989).

While FLAIR is a good indicator of injury to the white matter, it can be over interpreted, and should not be the sole determinant of an injury signal. Other MR methods, including MR spectroscopy and DTI are more reliable measures of pathological changes in white matter. MR spectroscopy showed injured white matter as a decrease in the NAA signal in vascular dementia as opposed to Alzheimer's disease (Kattapong et al., 1996). Patients with vascular dementia of the Binswanger type had reduced NAA in ^{1}H-MRS as opposed to those with AD and age-matched controls. The ratios of *N*-acetyl groups to creatine and *N*-acetyl groups to choline were lower in regions of subcortical white matter in patients with vascular dementia than in patients with dementia of the Alzheimer type. Proton magnetic resonance spectroscopy is useful in distinguishing between injured and normal white matter (Brooks et al., 1997). ^{1}H-MRS spectroscopic measurements of white matter injury provide an excellent indicator of white matter damage. NAA levels correlate better with neuropsychological measures of executive function than WMH volume. The correlation is good for both NAA and creatinine (CR), another molecule that can be used to indicate white matter damage (Gasparovic et al., 2013). However, the complexity of obtaining ^{1}H-MRS measurements, particularly on clinical scanners, has limited their use clinically. Since diffusion methods can be done on clinical scanners, they have replaced spectroscopy as the method of choice to show white matter injury.

Diffusion tensor imaging

Diffusor tensor imaging (DTI) is important when a tissue—such as the neural axons of white matter in the brain—has an internal fibrous structure analogous to the anisotropy of some crystals. Water will then diffuse more rapidly in the direction aligned with the internal structure, and more slowly as it moves perpendicular to the preferred direction. This also means that the measured rate of diffusion will differ depending on the direction from which an observer is looking. Traditionally, in diffusion-weighted imaging, three gradient-directions are applied, sufficient to estimate the trace of the diffusion tensor or "average diffusivity," a putative measure of edema. Clinically, trace-weighted images have proven to be very useful to diagnose vascular strokes in the brain, early detection (within a couple of minutes) of hypoxic edema. More extended DTI scans derive neural tract directional information from the data using 3D or multidimensional vector algorithms based on six or more gradient directions, sufficient to compute diffusion tensors. The diffusion tensor model is a rather simple model of the diffusion process, assuming homogeneity and linearity of the diffusion within each image voxel. From the diffusion tensor, diffusion anisotropy measures such as the fractional anisotropy (FA), can be computed. Moreover, the principal direction of the diffusion tensor can be used to infer the white-matter connectivity of the brain (i.e., tractography; trying to see which part of the brain is connected to which other part). Recently, more advanced models of the diffusion process have been proposed that aim to overcome the weaknesses of the diffusion tensor model. Amongst others, these include q-space imaging and generalized DTI.

In present-day clinical neurology, various brain pathologies may be best detected by looking at particular measures of anisotropy and diffusivity. The underlying physical process of diffusion causes a group of water molecules to move out from a central point, and gradually reach the surface of an ellipsoid if the medium is anisotropic (that would be the surface of a sphere for an isotropic medium). The ellipsoid formalism functions also as a mathematical method of organizing tensor data. Measurement of an ellipsoid tensor further permits a retrospective analysis, to gather information about the process of diffusion in each voxel of the tissue. In an isotropic medium such as cerebrospinal fluid, water molecules move due to diffusion and they move at equal rates in all directions. By knowing the detailed effects of diffusion gradients, we can generate a formula that allows us to convert the signal attenuation of an MRI voxel into a numerical measure of diffusion—the diffusion coefficient *D*. When various barriers and restricting factors such as cell membranes and microtubules interfere with the free diffusion, we are measuring an "apparent diffusion coefficient," or ADC, because the measurement misses all the local effects and treats the attenuation as if all the movement rates were solely due to Brownian motion. The ADC in anisotropic tissue varies depending on the direction in which it is measured. Diffusion is fast along the length of (parallel to) an axon, and slower perpendicularly across it. Once we have measured the voxel from six or more directions and corrected for attenuation due to T2 and T1 effects, we can use information from the calculated ellipsoid tensor to describe what is happening in the voxel. If you consider an ellipsoid sitting at an angle in a Cartesian grid, then you can consider the projection of that ellipse onto the three axes. The three projections can give you the ADC along each of the three axes: ADC_x, ADC_y, ADC_z.

The diffusivity along the principal axis, λ_1 is also called the longitudinal diffusivity or the axial diffusivity or even the parallel diffusivity $\lambda_{\parallel}$. The diffusivities in the two minor axes are often averaged to produce a measure of radial diffusivity (Fig. 2.7) (Raja et al., 2019).

DTI data can be used to perform tractography within white matter. Fiber-tracking algorithms can be used to track a fiber along its whole length (e.g., the corticospinal tract, through which the motor information transits from the motor cortex to the spinal cord and the peripheral nerves). Tractography is a useful tool for measuring deficits in white matter, such as in aging. Its estimation of fiber orientation and strength is increasingly accurate, and it has widespread potential implications in the fields of cognitive neuroscience.

The use of DTI has been generalized for use in clinical studies by using white matter masks and the creation of skeletons of the deeper white matter that can be used to perform diffusion calculations, which can be repeated by multiple investigators and used for research studies. The goal is to establish a fully automated, robust imaging marker for cerebral SVD and related cognitive impairment that is easy to implement, reflects disease burden, and is strongly associated with processing speed, the predominantly affected cognitive domain in SVD. To accomplish this, a novel MRI marker was created

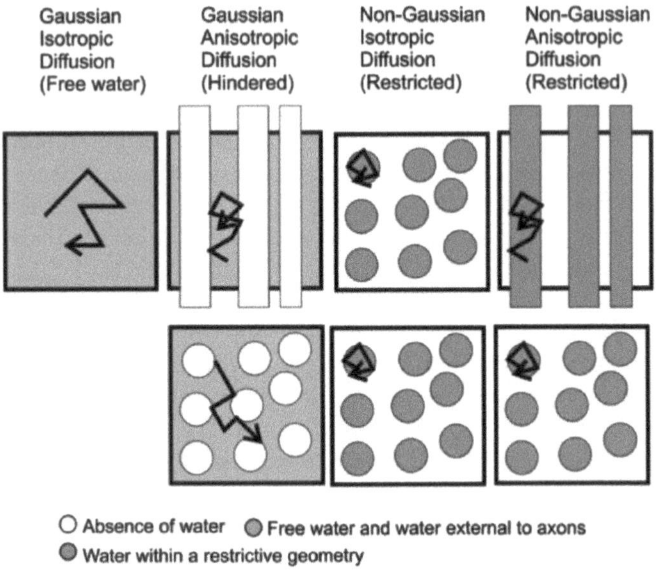

FIGURE 2.7 The figure shows examples of when the diffusion is Gaussian and non-Gaussian and when it is isotropic and anisotropic. Isotropic diffusion occurs in fluids without barriers or spherically symmetric barriers. Anisotropic diffusion occurs when there is a preferential direction for diffusion. In a geometry with a bundle of cylindrical tubes, diffusion in the fluid outside the tubes and the diffusion of water in the tubes are both anisotropic because the diffusion along the length of the tubes is greater in the direction of the tube's axis. The diffusion in the space outside the tubes can be approximated by Gaussian diffusion while the diffusion within the tubes is bounded by the tube walls and is non-Gaussian. The diffusion in the space external to the tubes is also hindered and the diffusion within the tubes is restricted diffusion.

based on DTI, namely, skeletonized white matter tracts, and histogram analysis. The marker [peak width of skeletonized mean diffusivity (PSMD)] was compared to conventional SVD imaging markers. PSMD was associated with processing speed in all study samples with SVD. PSMD explained most of the variance in processing speed and consistently outperformed conventional imaging markers (WMH volume, lacune volume, and brain volume) in multiple regression analyses. Increases in PSMD were linked to vascular, but not to neurodegenerative, disease. In longitudinal analysis, PSMD captured SVD progression better than other imaging markers (Baykara et al., 2016).

DTI differentiated patients with SIVD from those with AD and correlated executive function during the early stage of disease; there were clearly different DTI profiles between the SIVD and AD groups on the basis of comparable disease severity and general cognition. Compared to the cognitive assessments, it was easier to use changes in FA and MD to differentiate SIVD from AD, which demonstrated different microstructural changes related to pathogenesis in these two diseases (Tu et al., 2017). Tractography shows the tracts connecting different regions of the brain.

Advanced diffusion techniques: fiber tract connectivity with mean kurtosis and constrained spherical deconvolution

There are several methods that provide a more precise exploration of the white matter microstructure that will be useful in differentiating different pathological processes. Each of these methods tries to more carefully model the complexity of the crossing fiber tracts. The diffusion kurtosis imaging (DKI) technique is an extension of DTI, with the diffusion being non-Gaussian. Mean kurtosis (MK), axial kurtosis (AK), and radial kurtosis (RK) are the main DKI derived metrics. Though DKI parametric maps look similar to FA maps in terms of intensity patterns, DKI outperforms DTI with higher sensitivity. Another diffusion model is the constrained spherical deconvolution (CSD) model which fits the diffusion weighted signal defined as a convolution product of a fiber orientation distribution function and a single response function of a coherent fiber; this model has the advantage of defining multiple fiber populations with different directions within a voxel. Fiber tracts can be calculated based on the direction of the fiber orientation distribution peaks. Fiber specific metrics such as track density imaging and average pathlength maps can be calculated based on a CSD model (Raja et al., 2020).

Detection of white matter injury

With the many imaging methods of identifying regions of white matter injury, it is important to understand the benefits and limitations of each one. CT was the first to be used: it provided a limited image of the regions of white matter injury as attenuation of the X-rays. MRI was much more sensitive with very clear pictures of regions of white matter changes, particularly in the FLAIR sequences. Increasing the sensitivity had both benefits and drawbacks, since it showed hyperintensities in healthy individuals, limiting its usefulness in disease detection. The presence of normal changes of aging like WMH is referred to as leukoaraiosis: in Greek, *leuko* means white and *araiosis* means pale; this was not intended to be used as a pathological term, but as a radiological finding (Hachinski et al., 1987). Other imaging methods more accurately identify abnormal white matter, including ^{1}H-MRS and diffusion methods, which are available on most MRI scanners, making them the methods of choice. As described above, there are a number of diffusion methods that can be used with a range of sensitivity. Since they are highly correlated, the method selected is not critical. The skeletonization procedure has been adopted for use in many research studies: it provides a single number indicating the degree of injury and is an improvement over the FLAIR method in showing actual injury. The newer methods involving DTI have the advantage that they can demonstrate specific fiber tracts that are involved, which is primarily of benefit for research studies. Diffusion is important in classifying patients with dual pathology of AD and vascular dementia (mixed dementia). This is an important distinction in planning treatment trials; detecting white matter injury with MRI methods in patients with a biological diagnosis of AD based on CSF proteins and PET studies is the optimal way to use biomarkers for MX patients.

References

Baykara, E., Gesierich, B., Adam, R., Tuladhar, A. M., Biesbroek, J. M., Koek, H. L., Ropele, S., Jouvent, E., Chabriat, H., Ertl-Wagner, B., Ewers, M., Schmidt, R., de Leeuw, F.-E., Biessels, G. J., Dichgans, M., & Duering, M. (2016). A novel imaging marker for small vessel disease based on skeletonization of white matter tracts and diffusion histograms. *Annals of Neurology*, *80*(4), 581–592. Available from https://doi.org/10.1002/ana.24758.

Binswanger's encephalopathy. (1981). Lancet (London, England), *1*(8226), 923.

Bitar, R., Leung, G., Perng, R., Tadros, S., Moody, A. R., Sarrazin, J., McGregor, C., Christakis, M., Symons, S., Nelson, A., & Roberts, T. P. (n.d.). MR pulse sequences: What every radiologist wants to know but is afraid to ask. *RadioGraphics, 26*(2), 513–537.

Brainard, J. R., Kyner, E., & Rosenberg, G. A. (1989). 13C nuclear magnetic resonance evidence for gamma-aminobutyric acid formation via pyruvate carboxylase in rat brain: a metabolic basis for compartmentation. *Journal of Neurochemistry*, *53*(4), 1285–1292.

Brooks, W. M., Wesley, M. H., Kodituwakku, P. W., Garry, P. J., & Rosenberg, G. A. (1997). 1H-MRS differentiates white matter hyperintensities in subcortical arteriosclerotic encephalopathy from those in normal elderly. *Stroke; A Journal of Cerebral Circulation*, *28*(10), 1940–1943.

Caplan, L. R., & Schoene, W. C. (1978). Clinical features of subcortical arteriosclerotic encephalopathy (Binswanger disease). *Neurology*, *28*(12), 1206–1215.

Chui, H. C., Victoroff, J. I., Margolin, D., Jagust, W., Shankle, R., & Katzman, R. (1992). Criteria for the diagnosis of ischemic vascular dementia proposed by the State of California Alzheimer's Disease Diagnostic and Treatment Centers. *Neurology*, *42*(3 Pt 1), 473–480.

Gasparovic, C., Prestopnik, J., Thompson, J., Taheri, S., Huisa, B., Schrader, R., Adair, J. C., & Rosenberg, G. A. (2013). 1H-MR spectroscopy metabolite levels correlate with executive function in vascular cognitive impairment. *Journal of Neurology, Neurosurgery, and Psychiatry*, *84*(7), 715–721. Available from https://doi.org/10.1136/jnnp-2012-303878.

Hachinski, V., Iadecola, C., Petersen, R. C., Breteler, M. M., Nyenhuis, D. L., Black, S. E., Powers, W. J., DeCarli, C., Merino, J. G., Kalaria, R. N., Vinters, H. V., Holtzman, D. M., Rosenberg, G. A., Wallin, A., Dichgans, M., Marler, J. R., & Leblanc, G. G. (2006). National Institute of Neurological Disorders and Stroke-Canadian Stroke Network vascular cognitive impairment harmonization standards. *Stroke; a Journal of Cerebral Circulation*, *37*(9), 2220–2241.

Hachinski, V. C., Iliff, L. D., Zilhka, E., Du Boulay, G. H., McAllister, V. L., Marshall, J., Russell, R. W., & Symon, L. (1975). Cerebral blood flow in dementia. *Archives of Neurology*, *32*(9), 632–637.

Hachinski, V. C., Potter, P., & Merskey, H. (1987). Leuko-araiosis. *Archives of Neurology*, *44*(1), 21–23.

Hoult, D. I., Busby, S. J., Gadian, D. G., Radda, G. K., Richards, R. E., & Seeley, P. J. (1974). Observation of tissue metabolites using 31P nuclear magnetic resonance. *Nature*, *252*(5481), 285–287.

Hunt, A. L., Orrison, W. W., Yeo, R. A., Haaland, K. Y., Rhyne, R. L., Garry, P. J., & Rosenberg, G. A. (1989). Clinical significance of MRI white matter lesions in the elderly. *Neurology*, *39*(11), 1470–1474.

Jack, C. R., Bennett, D. A., Blennow, K., Carrillo, M. C., Dunn, B., Haeberlein, S. B., Holtzman, D. M., Jagust, W., Jessen, F., Karlawish, J., Liu, E., Molinuevo, J. L., Montine, T., Phelps, C., Rankin, K. P., Rowe, C. C., Scheltens, P., Siemers, E., Snyder, H. M., & Sperling, R. (2018). NIA-AA Research Framework: Toward a biological definition of Alzheimer's disease. *Alzheimer's & Dementia: The Journal of the Alzheimer's Association*, *14*(4), 535–562. Available from https://doi.org/10.1016/j.jalz.2018.02.018.

Kattapong, V. J., Brooks, W. M., Wesley, M. H., Kodituwakku, P. W., & Rosenberg, G. A. (1996). Proton magnetic resonance spectroscopy of vascular- and Alzheimer-type dementia. *Archives of Neurology*, *53*(7), 678–680.

Klunk, W. E., Wang, Y., Huang, G. F., Debnath, M. L., Holt, D. P., & Mathis, C. A. (2001). Uncharged thioflavin-T derivatives bind to amyloid-beta protein with high affinity and readily enter the brain. *Life Sciences*, *69*(13), 1471–1484.

McKhann, G., Drachman, D., Folstein, M., Katzman, R., Price, D., & Stadlan, E. M. (1984). Clinical diagnosis of Alzheimer's disease: Report of the NINCDS-ADRDA Work Group under the auspices of Department of Health and Human Services Task Force on Alzheimer's Disease. *Neurology*, *34*(7), 939–944.

Raja, R., Caprihan, A., Rosenberg, G. A., Rachakonda, S., & Calhoun, V. D. (2020). Discriminating VCID subgroups: A diffusion MRI multi-model fusion approach. *Journal of Neuroscience Methods*, *335*, 108598. Available from https://doi.org/10.1016/j.jneumeth.2020.108598.

Raja, R., Rosenberg, G., & Caprihan, A. (2019). Review of diffusion MRI studies in chronic white matter diseases. *Neuroscience Letters*, *694*, 198–207. Available from https://doi.org/10.1016/j.neulet.2018.12.007.

Rosenberg, G. A., Kornfeld, M., Stovring, J., & Bicknell, J. M. (1979). Subcortical arteriosclerotic encephalopathy (Binswanger): Computerized tomography. *Neurology*, *29*(8), 1102–1106.

Sokoloff, L. (1984). Modeling metabolic processes in the brain in vivo. *Annals of Neurology*, *15*, S1–S11, Suppl.

Tu, M.-C., Lo, C.-P., Huang, C.-F., Hsu, Y.-H., Huang, W.-H., Deng, J. F., & Lee, Y.-C. (2017). Effectiveness of diffusion tensor imaging in differentiating early-stage subcortical ischemic vascular disease, Alzheimer's disease and normal ageing. *PLoS One*, *12*(4), e0175143. Available from https://doi.org/10.1371/journal.pone.0175143.

Chapter 3

Biomarkers in cerebrospinal fluid and blood aid diagnosis of dementia

Pathologic deposition of misfolded proteins in the brain: amyloid, tau, α-synuclein, TDP-43

The neuropathology of dementia was established in the early 1900s by the discoveries from two neuroscientists, Camillo Golgi and Santiago Ramon y Cajal. Golgi developed staining methods that Cajal used as the basis for his theories of the connectivity of cells that have been shown to be correct. For their pioneering studies of cells in the brain, each man received the Nobel Prize. Using the new silver staining method, Alois Alzheimer identified plaques and tangles in the brain of a severely demented woman who died in her fifties. Another neuropathologist, Oskar Fisher, working around the same time as Alzheimer, described a large series of demented patients with "plaques" similar to those seen by Alzheimer (Goedert, 2009). Because of a lack of interest in diseases of old age, only a few pathological studies were reported on patients with dementia until the 1970s. The disease described by Alzheimer was considered rare since the normal lifespan was less than 60 years of age and vascular disease of the brain was considered the most common.

This changed dramatically when the neurologist Robert Katzman, working with neuropathologist Robert Terry, proposed that death certificates were under reporting the numbers of patients dying with dementia, particularly Alzheimer's disease (AD) (Terry & Katzman, 1983). Recognizing that this was a national problem with people living longer, the National Institutes of Health established the first Alzheimer's Disease Research Center (ADRC) led by Dr. Katzman in Bronx, New York; these have now grown to 34 and have become the major sites of dementia research. In the 1970s, several other significant advances catapulted dementia from obscurity into one of the major neurological disorders. The first advance was the invention of computed tomography and the rapid expansion of the tomographic methods into magnetic resonance imaging (MRI) and positron emission tomography (PET),

Neuroinflammation in Vascular Dementia. DOI: https://doi.org/10.1016/B978-0-12-823455-6.00001-8

initiating the modern era of neuroimaging in dementia research. The second breakthrough was in biochemistry with the discovery by George Glenner and his colleagues that the plaques observed by Alzheimer contained amyloid, a common protein present in many organs (Glenner & Wong, 1984).

Neuroimaging revived interest in pathological studies to confirm the findings seen during life, and along with the increased longevity due to improvements in medicine, produced advances in understanding of the underlying disease mechanisms. For many years, the discovery of amyloid in plaques and subsequently in cerebrospinal fluid (CSF) formed the basis for the dominant theory of AD, the "amyloid hypothesis" (Musiek & Holtzman, 2015). After three decades, the amyloid hypothesis in dementia research resulted in the failure of several large studies of agents to remove amyloid from the brain to improve cognition. This has resulted in a reassessment of the unitary amyloid theory and the emergence of alternative hypotheses based on the other misfolded proteins associated with dementia, and the role of vascular disease and neuroinflammation in accelerating cognitive decline. Ironically, the recent increase in pathological studies has identified vascular disease as a key disease process in the aging brain, returning to the original concepts of the role of vascular disease in dementia. This has led to the realization that the combination of misfolded proteins with vascular disease or mixed dementia, is the most common form of the disease in the elderly. The current thinking is that while amyloid is important in the pathophysiology of the early stages of AD, it is one of several misfolded proteins that accumulate in the brain with aging, and that these proteins trigger an inflammatory response that accelerates the cognitive failure. Four misfolded proteins are found deposited in the brains of cognitively impaired patients; it is likely that others will be discovered. These include amyloid and tau as well as α-synuclein and transactive response DNA-binding protein 43 kDa (TDP-43) (Karanth et al., 2020).

The role of amyloid led to the development of transgenic mouse models that overexpressed human amyloid genes; these models have provided important information on the pathologic role of amyloid and they have been used to test various therapeutic agents to reduce the amyloid load in the brain. A number of antibodies against different components of the amyloid molecules were developed and tested in transgenic animal models. Although they successfully reduced amyloid in the brains of animals and humans, and have led to clinical trials, to date they have shown disappointing results. The failure of the anti-amyloid treatments has led to the development of another antibody approach that uses agents that reduce tau. More importantly, it has stimulated research that is not based on the amyloid hypothesis that is focused on the inflammatory response triggered by the abnormal proteins in the brain (Selkoe, 2019). Despite the failure of anti-amyloid treatments, there is considerable evidence that it plays a role in the disease, most likely in the initiation of the disease.

Amyloid is a complex molecule and the consequences of amyloid in different locations in the brain are poorly understood. It can be found inside the brain tissue in the amyloid plaques that form around intracerebral blood vessels and in the blood vessels of the meninges close to the surface. Although the amyloid in both locations is similar, that found around blood vessels is more destructive because of the prevalence of intracerebral hemorrhage in patients with amyloid angiopathy. A similar predilection to bleed produces microbleeds in the brain. MRI shows these microbleeds with special sequences that increase the susceptibility to iron deposited around the vessels. Families with congenital cerebral amyloid angiopathy (CAA) are prone to intracerebral hemorrhage; cerebral microbleeds are less dramatic examples of the consequences of amyloid being deposited in blood vessels (Greenberg et al., 2020).

The amyloid in the brain is produced from the amyloid precursor protein (APP) by a series of cleavages by secretases, creating a continuous production of amyloid as soluble and insoluble fragments. This continual production requires a parallel mechanism for removal. A growing body of evidence implicates the removal of the amyloid via flow of interstitial fluid along perivascular pathways (Weller et al., 2008). CAA might also contribute to the pathogenesis of AD by slowing perivascular drainage, leading to its buildup in the brain. Given that overlap between sporadic CAA and AD is high, virtually every clinical trial of treatment for sporadic AD or CAA can be considered as a trial for treatment of both, underlining the importance of identifying the interactions between these two processes in disease pathogenesis.

Hemorrhage is a major consequence of CAA, ranging from small microscopic bleeds to large intracerebral life-threatening hemorrhages, which may be due to the action of amyloid to induce an inflammatory response, which leads to the synthesis, release, and activation of the blood–brain barrier (BBB) disrupting matrix metalloproteinases (MMPs). A study of murine cerebral endothelial cells showed that release of MMP-9 increased bleeding, resulting in increased extracellular matrix degradation. Aged APP transgenic mice, a mouse model used to study AD, had increased intracerebral hemorrhage. Furthermore, extensive MMP-9 immunoreactivity was observed in CAA vessels with evidence of microhemorrhage, but not detected in aged wild type or young APP mice. MMP-9 expression, stimulated by amyloid-β (Aβ), may play a role in the pathogenesis of spontaneous intracerebral hemorrhage in patients with CAA (Lee et al., 2003).

Biomarkers for amyloid in cerebrospinal fluid

After the identification of amyloid as the abnormal protein in the plaques, a search began to identify amyloid in the CSF to provide a biomarker to improve diagnosis. Multiple fragments of the amyloid molecule derived from the APP were identified; these variants of the beta amyloid protein, differing only at their carboxy terminus (beta 1–39, beta 1–40, beta 1–42, and beta 1–43), were thought to be the major components of the cerebral amyloid deposits. Kinetic studies of aggregation by three naturally occurring beta protein variants (beta 1−39, beta 1−40, beta 1−42) and four model peptides (beta 26−39, beta 26−40, beta 26−42, beta 26−43) demonstrated that amyloid formation, like crystallization, is a nucleation-dependent phenomenon (Jarrett et al., 1993).

The discovery of the fragments of the APP paved the way for the development of assays to measure them in the CSF, particularly the β_{1-42} and β_{1-40} fragments (Blennow et al., 2010). Biomarkers for amyloid opened an important area of investigation to the earliest biomarkers of AD with the goal of initiating disease-modifying therapy before the disease progresses beyond the point where treatment would be ineffective. Current clinical diagnostic criteria for AD require a patient to have dementia before a diagnosis can be made, and are largely based on the exclusion of other disorders (McKhann et al., 1984). Disease-modifying drugs for AD, when they become available, will need to be administered very early in the course of the disease, before neurodegeneration is too severe and cellular damage is widespread. No clinical method is available for identifying prodromal AD in patients with mild cognitive impairment (MCI); these individuals have only early disturbances in episodic memory. The CSF biomarkers total tau (tTau), phosphorylated tau (p-tau$_{181}$ and p-tau$_{231}$) and β-amyloid$_{1-42}$ have a high diagnostic accuracy for AD, and for prodromal AD in patients with MCI.

Amyloid and tau

There is considerable evidence in support of the amyloid hypothesis, such as the development of AD in patients with Down's syndrome related to a genetic defect in the APP, the early onset of AD in patients with specific genes causing an increase in amyloid in the brain up to 20 years to development, and the ability of producing AD-like pathology in transgenic mice with the amyloid gene. The failure of a number of clinical trials with antibodies to AD has been disappointing and has resulted in alternate hypotheses that can be tested in clinical trials (Hardy, 2017; Hardy & Selkoe, 2002). Several promising areas of research are related to the induction of an inflammatory response secondary to the deposition of amyloid. The major cell type involved in the inflammatory response is the microglia, which are similar to the macrophages in the systemic circulation; they are involved in innate immunity. While the search for alternative theories of AD pathogenesis is being pursued, another path to treat AD is based on the realization that the failure of the early amyloid antibody trials may have been related to the disease being too far advanced to be amenable to treatment. This has led to an intense search for biomarkers to predict the presence of the disease at an earlier stage; the biomarkers were sought in neuroimaging, CSF, and blood. In addition, a parallel process has been pursued to use clinical and neuropsychological examination to identify the early stages of the disease.

Neuroimaging biomarkers include generalized and focal atrophy with emphasis on cortical loss and hippocampal atrophy. While this is apparent in many patients at the earliest stages of AD, brain volume loss in patients with dementia overlaps with normal elderly, making it less useful as a biomarker. A more promising approach is the measurement of various forms of amyloid and tau in CSF and more recently in plasma. Much work has been done on the role of CSF biomarkers in the diagnosis of early AD. Amyloid has several forms that can be measured, including $A\beta_{1-42}$, $A\beta_{1-38}$, $A\beta_{1-40}$. The ratio of $A\beta_{1-42}/A\beta_{1-40}$ is more accurate than either alone. Changes in CSF amyloid can be seen as much as 20 years prior to onset of AD in patients with one of the genes that cause familial AD. Changes in tTau and p-Tau are linked to changes in Aβ, which begin to accumulate in plaques first. There are a number of phosphorylation sites in the tau protein with the two most carefully studied being at 181 and 217. In a large cohort of early onset AD with familial dominant genes, there was a pattern of tau phosphorylation that suggested that Aβ needed to be present. It is possible to estimate the year of onset in the Dominantly Inherited Alzheimer Network (DIAN) studies of early onset of AD in families. The earliest changes in tau phosphorylation were seen at position 217, followed by 181 (Barthélemy, Li, et al., 2020). Patterns of tau phosphorylation in the CSF of DIAN mutation carriers vary over the course of AD progression. This begins once aggregated Aβ pathology measured by PiB−PET is established decades before symptoms, and subsequently unfolds over a period of nearly two decades. This suggests that quantifying soluble tau phosphorylation occupancy can track the AD process across its preclinical to symptomatic stages, and confirm the link between Aβ pathology and tau hyperphosphorylation. Models have been made to show the progression from amyloid through tau deposition, leading to cognitive impairment; these models should be considered as indicative of the knowledge at the time (Fig. 3.1) (Jack & Holtzman, 2013).

Biomarkers in mild cognitive impairment

The need to identify patients with AD earlier in the course has led to the introduction of the concept of MCI. MCI is classified into two subtypes: amnestic and nonamnestic. Amnestic MCI is clinically significant memory impairment that does not meet the criteria for dementia. Typically, patients and their families are aware of increasing forgetfulness. However, other cognitive capacities, such as executive functions, use of language, and visuospatial skills, are relatively preserved, and daily activities are intact, except perhaps for some mild inefficiencies. Nonamnestic MCI is characterized by a subtle decline in functions not related to memory, affecting attention, use of language, or visuospatial skills. The nonamnestic type of MCI is probably less common than the amnestic type and may be the forerunner of dementias that are not related to AD, such as frontotemporal lobar degeneration or

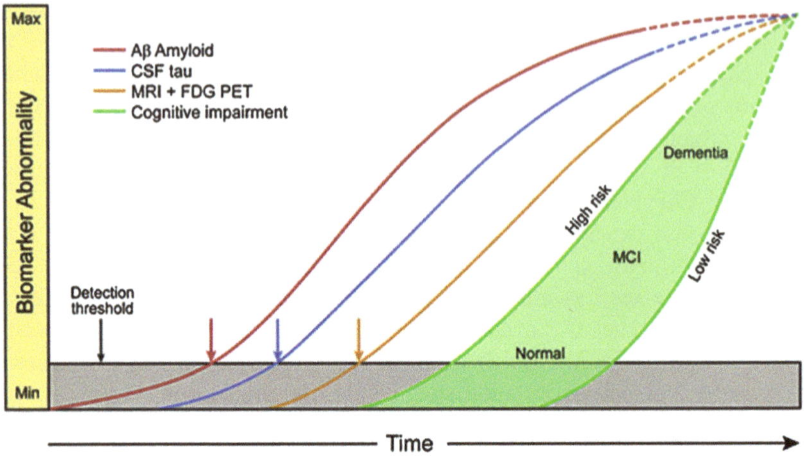

FIGURE 3.1 **Biomarker model of pure AD.** The horizontal axis is time and the vertical axis severity of biomarker abnormality from completely normal (min) to abnormal (max). The threshold for biomarker detection of pathophysiology is denoted by a horizontal line. The gray area denotes the zone in which abnormal pathophysiology lies below the biomarker detection threshold. Amyloid biomarkers become abnormal first, followed by CSF tau, followed by fluorodeoxyglucose (FDG) PET and MRI. Cognitive impairment (*green filled area*) is the last event in the progression of the disease. A range of cognitive responses are possible that depend on the individual's risk profile. The cognitive response curve is shifted to the left for those with low cognitive reserve and to the right for those with high cognitive reserve. At a given point on the disease time line, a person with high cognitive reserve can be cognitively normal while a low cognitive response person with the same biomarker profile can be impaired. All biomarker curves (as well as cognitive impairment) are configured as sigmoids, but the curves have a progressively steeper slope in the right-hand tail for later-changing biomarkers. The right-hand tails of the curves are dashed, indicating that biomarker trajectories in end stage dementia are unknown at this time. Alzheimer's disease; *CSF*, cerebrospinal fluid; *MCI*, mild cognitive impairment; *MRI*, magnetic resonance imaging; *PET*, positron emission tomography. *From Jack, C. R. & Holtzman, D. M. (2013). Biomarker modeling of Alzheimer's disease.* Neuron, 80*(6), 1347–1358. https://doi.org/10.1016/j.neuron.2013.12.003.*

dementia with Lewy bodies (Petersen, 2016). Earlier signs of memory impairment can be experienced by elderly individuals prior to the ability of cognitive testing to detect them. These early signs have been called subjective memory decline. A substantial group of participants (14% in one large study) with subjective memory complaints were at increased risk of incident MCI based on combined baseline endorsement of any consistent subjective cognitive disorder (SCD) on several memory scales, such as the Everyday Cognitive scale (van Harten et al., 2018).

Differentiating MCI from normal aging is challenging. It is more than subtle forgetfulness; cognitive impairment is more prominent in persons with mild amnestic cognitive impairment. All other aspects of behavior are preserved, and the extent of memory decline is most apparent to others. Some patients with MCI progress to AD while others remain stable for many years.

As the disease process progresses, there is a gradual loss of other cognitive functions, and a more pronounced problem retaining information and finding words. Disorientation results in difficulty with orientation when driving and getting lost when walking in the neighborhood. Eventually, it becomes obvious that there is a major problem with cognition as the language becomes sparse, personal hygiene declines, and personality changes cause angry outbursts with paranoid thinking. Population-based studies of MCI estimate its prevalence to be between 15% and 20% in persons 60 years of age and older, making it a common condition. The annual rate at which MCI progresses to dementia varies between 8% and 15% per year. In those MCI cases destined to develop AD, biomarkers are emerging to help identify etiology and predict progression. However, not all MCI is due to AD, and identifying subtypes is important for possible treatment and counseling. If treatable causes are identified, the person with MCI might improve (Petersen, 2016).

Progressively, AD affects most areas of the brain: interfering with memory, thinking, judgment, language, problem-solving, personality and, eventually, movement. There are five stages associated with AD: (1) preclinical AD, (2) MCI, (3) mild dementia due to AD, (4) moderate dementia due to AD, and (5) severe dementia due to AD. Dementia is used to describe a group of symptoms that affect intellectual and social abilities severely enough to interfere with daily function. Although discussed as stages, the disease is a continuous process with preclinical AD beginning long before symptoms are noted; at this stage examination of the CSF can detect abnormal proteins associated with AD – Aβ and p-Tau. Where available, PET with isotopic specific ligands that bind to amyloid and tau is useful, but too expensive for routine use. Generalized decline in metabolism can be observed with fluorodeoxyglucose (FDG)-PET. Genetic tests for apolipoprotein E (APOE) genes can identify those at higher risk for AD, particularly late-onset AD. MCI, as the name implies, describes mild changes in memory and cognition; this is a frequently used diagnosis, but suffers from a lack of clarity in its definition, making it difficult to determine exactly how best to use it. Use of the term is further complicated by the fact that many with the diagnosis will fail to progress to AD since not everyone with MCI has AD. As the disease progresses, individuals have difficulty remembering newly learned information and ask the same question over and over. Those that do progress show increasing difficulty solving problems, doing complex tasks and making sound judgments. Planning a family event or balancing a checkbook may become overwhelming. Many people experience lapses in judgment and may experience a change in personality. Apathy develops particularly in novel social situations; they may show uncharacteristic irritability or anger. Other symptoms in this moderate stage include reduced motivation, difficulty organizing and expressing thoughts. Individuals have increasing trouble finding their way around, even in familiar places. During the moderate dementia stage of AD, people grow more confused and forgetful and begin to need more help with daily activities and self-care. People with the moderate dementia stage of Alzheimer's disease may show increasingly poor judgment and deepening confusion. Individuals lose track of where they are, the day of the week, or the season. They may confuse family members or close friends with one another or mistake strangers for family. They may wander, possibly in search of surroundings that feel more familiar. These difficulties make it unsafe to leave those in the moderate dementia stage on their own. As they experience even greater memory loss, patients may forget details of their personal history, such as their address or phone number, or where they attended school. They repeat favorite stories or make up stories to fill gaps in memory. At this stage they need help with some daily activities. Assistance may be required with choosing proper clothing for the occasion or the weather and with bathing, grooming, using the bathroom and other self-care. Some individuals occasionally lose control of their bladder or bowel movements. Undergo significant changes in personality and behavior. It's not unusual for people with the moderate dementia stage to develop unfounded suspicions – for example, to become convinced that friends, family, or professional caregivers are stealing from them or that a spouse is having an affair. Others may see or hear things that aren't really there. Individuals often grow restless or agitated, especially late in the day. Some people may have outbursts of aggressive physical behavior. Severe dementia due to AD in the late stage of the disease includes continuing decline in mental function and begins to have an impact on movement and physical abilities. In late stage severe dementia due to AD, people generally lose the ability to communicate coherently. An individual can no longer converse or speak in ways that make sense, although he or she may occasionally say words or phrases. They require daily assistance with personal care, including total assistance with eating, dressing, using the bathroom, and all other daily self-care tasks. As they experience a decline in physical abilities, they lose the ability to walk without assistance, reaching the point where they are unable to sit or hold up their head without support. In the final stages, muscles may become rigid and reflexes abnormal. Eventually, a person loses the ability to swallow and to control bladder and bowel functions. The rate of progression through AD stages varies widely. On average, people with AD live between 3 and 11 years after diagnosis, but some survive 20 years or more. The degree of impairment at diagnosis can affect life expectancy. Pneumonia is a common cause of death because impaired swallowing allows food or beverages to enter the lungs, where an infection can begin. Other common causes of death include dehydration, malnutrition, falls, and other infections.

Clinical and biological diagnosis of Alzheimer's disease

Diagnosis of AD was based on clinical criteria for many years without the use of ancillary testing (McKhann et al., 1984). Autopsy studies of large series of patients diagnosed clinically with AD showed a high incidence of other diagnoses, primarily vascular dementia. This was shown in community-based dementia studies (Schneider et al., 2007). In the ADRC autopsy database, patients with suspected AD based on clinical criteria were autopsied; those with suspected vascular disease were excluded. However, on review of the patients in the database, it was discovered that a large number of them had pathological evidence of vascular disease (Toledo et al., 2013). When vascular pathology was found, it was called mixed dementia, making that the largest group of patients; it is estimated that 70% of patients are in this category.

PET is another method that can detect amyloid and tau proteins present in the brain as well as regions of hypometabolism associated with atrophy. Ligands that bind to amyloid were the first to be developed, and they were compared to findings in the CSF, showing good agreement. Tau-binding ligands proved more difficult to develop, but several are now in routine use (Fig. 3.2) (Ashton et al., 2020).

Because of the cost of the equipment and the need for a large team of chemists and physicists to accurately measure the presence of these ligands in the brain, only a few research level centers are using amyloid and tau PET scans, relying on FDG-PET. Plasma measurements of AD proteins can be done with ultrasensitive single molecule array (SIMOA). Quantification of plasma p-tau_{231} has been done in four independent cohorts ($n = 588$) in different settings, including the full AD continuum and non-AD neurodegenerative disorders. Plasma p-tau_{231} was able to identify patients with AD and differentiate them from Aβ negative cognitively unimpaired older adults with high accuracy. Plasma p-tau_{231} also distinguished AD patients from patients with non-AD neurodegenerative disorders, as well as from Aβ negative MCI patients. In a neuropathology cohort, plasma p-tau_{231} in samples taken on average 4.2 years prior to postmortem very accurately identified AD neuropathology in comparison to non-AD neurodegenerative disorders. This is despite all patients being given an AD dementia diagnosis during life. Plasma p-tau_{231} was highly correlated with CSF p-tau_{231}, tau pathology as assessed by [^{18}F]MK-6240 PET, and brain amyloidosis by [^{18}F]AZD469 PET. Plasma p-tau_{231} is a promising novel biomarker of emerging AD pathology with the potential to facilitate clinical trials to identify vulnerable populations below PET threshold of Aβ positivity or apparent entorhinal tau deposition (Fig. 3.3) (Ashton et al., 2021).

In an attempt to improve diagnostic accuracy, particularly for research studies, a biological definition of AD has been proposed: this is based on the measurement of amyloid and tau proteins in the CSF or detection of them with PET scanning. The term "AD" in the NIA-AA research framework refers to an aggregate of neuropathologic changes and thus is defined in vivo by biomarkers and by postmortem examination, not by clinical symptoms alone (Jack et al., 2018). Amyloid (A), tau (T) and neurodegeration (N) or AT(N) use three groups of biomarkers based on the nature of the pathologic process

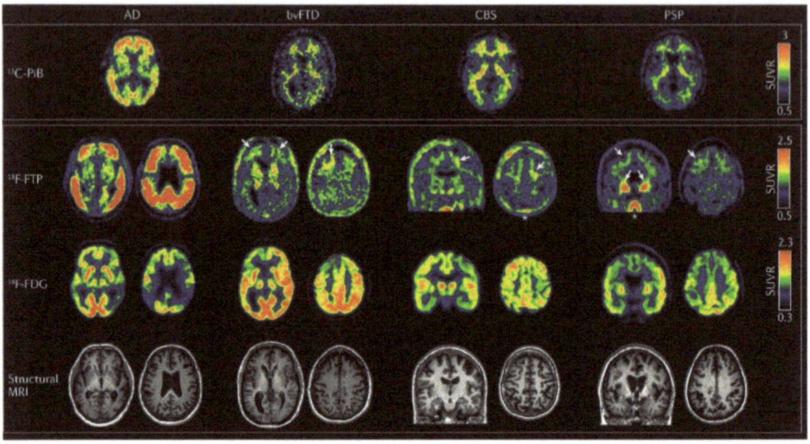

FIGURE 3.2 Multimodal imaging techniques. The top row shows axial slices of 11C-PiB scans, which reflect neuritic Aβ plaque density, from four patients. The second row shows PET images acquired using the tracer ^{18}F-FTP, which binds to intracellular aggregates of abnormally phosphorylated tau. Tracer binding in the individual with AD is high in temporoparietal areas, including the posterior cingulate and precuneus, as well as the dorsal prefrontal cortex. In the individuals with bvFTD, CBS or PSP, the arrowheads highlight areas of mild–moderate tracer binding. Asterisks indicate brain regions of nonspecific tracer retention ("off-target" binding), which overlap with known patterns of off-target binding in healthy individuals, including binding in the basal ganglia, midbrain regions, and cerebellum. The third row shows PET images acquired using the tracer ^{18}F-FDG, the retention of which is a marker for glucose uptake by the tissue. Across all individuals, regions of low ^{18}F-FDG retention overlap with regions of high ^{18}F-FTP retention. The bottom row shows structural T1-weighted MRI scans that correspond to the slices shown for the ^{18}F-FTP and ^{18}F-FDG scans. *11C-PiB*, 11C-Pittsburgh compound B; *^{18}F-FDG*, ^{18}F-fluorodeoxyglucose; *^{18}F-FTP*, ^{18}F-flortaucipir; *^{18}F-FTP*, ^{18}F-flortaucipir; *Aβ*, amyloid-β; *AD*, Alzheimer's disease; *bvFTD*, behavioral variant FTD; *CBS*, corticobasal syndrome; *MRI*, magnetic resonance imaging; *PET*, positron emission tomography; *PSP*, progressive supranuclear palsy. *From Ashton, N. J., Hye, A., Rajkumar, A. P., Leuzy, A., Snowden, S., Suárez-Calvet, M., Karikari, T. K., Schöll, M., La Joie, R., Rabinovici, G. D., Höglund, K., Ballard, C., Hortobágyi, T., Svenningsson, P., Blennow, K., Zetterberg, H. & Aarsland, D. (2020). An update on blood-based biomarkers for non-Alzheimer neurodegenerative disorders.* Nature Reviews Neurology, 16*(5), 265–284. https://doi.org/10.1038/s41582-020-0348-0.*

that each measures; these research criteria are not intended for routine clinical use. The goal of identifying biological biomarkers was to more objectively define and stage the disease primarily for use in research studies. The authors of the biological criteria realized that additional biomarkers would eventually be defined and complete the terminology; they specifically mentioned the inclusion of factors for vascular disease, α-synuclein, and TDP-43 as these biomarkers were discovered. Vascular disease is the most common of the other features; it can be determined using MRI to show the injury to the white matter, which is well developed for use in the definition of vascular cognitive impairment. Adding the vascular factor in combination with the AD measures can be done because the biomarkers are currently available.

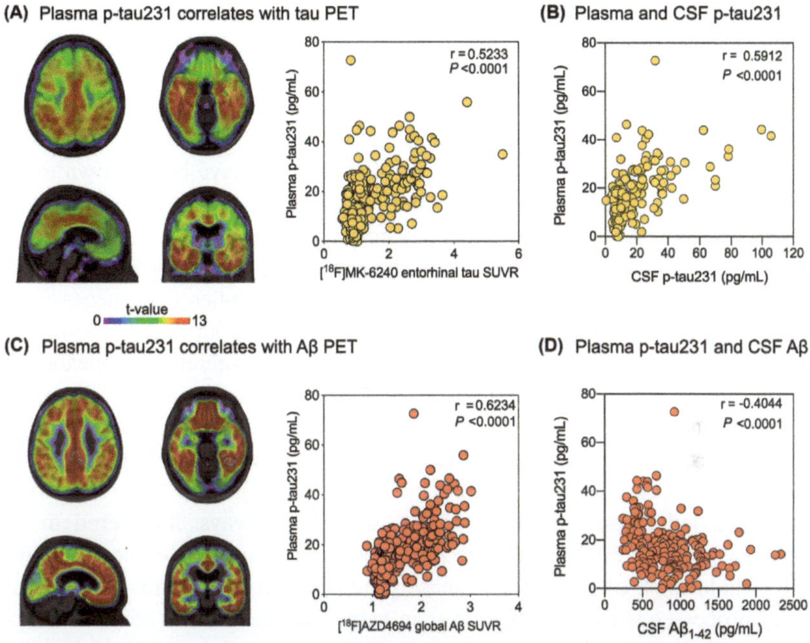

FIGURE 3.3 Plasma p-tau$_{231}$ concentration correlates with PET and CSF tau and Aβ. The brain maps show the results of linear regressions and a false discovery rate corrected for multiple comparisons at $P < 0.05$, whereas the plots show the results of Spearman rank correlation between biomarkers. (A) Plasma p-tau$_{231}$ concentrations were associated with [^{18}F]MK-6240 SUVR across the cortex with the highest association in the temporal and posterior cingulate cortices as well as with (B) CSF p-tau$_{231}$ concentrations. (C) Plasma p-tau$_{231}$ correlates with [^{18}F] AZD4694 SUVR across the cortex and with (D) CSF Aβ_{1-42} concentrations. *Aβ*, Amyloid-β *CSF*, cerebrospinal fluid; *PET*, positron emission tomography; *p-tau*, phosphorylated tau; *SUVR*, standardized uptake value ratio. *From Ashton, N. J., Pascoal, T. A., Karikari, T. K., Benedet, A. L., Lantero-Rodriguez, J., Brinkmalm, G., Snellman, A., Schöll, M., Troakes, C., Hye, A., Gauthier, S., Vanmechelen, E., Zetterberg, H., Rosa-Neto, P. & Blennow, K. (2021). Plasma p-tau231: A new biomarker for incipient Alzheimer's disease pathology.* Acta Neuropathologica, 141*(1–15), 1–16. https://doi.org/10.1007/s00401-021-02275-6.*

Theories of Alzheimer's disease: amyloid hypothesis

The isolation of amyloid from the plaques by George Glenner in 1984 was a major breakthrough that initiated a massive search for the link between amyloid and dementia. This discovery led to the dominance of the amyloid hypothesis in AD research that continues today (Raz et al., 2018). Early formulations of the amyloid hypothesis concentrated on the ability of the amyloid to form into fibrils and aggregate. The misfolded protein initiated the damage to tau forming the tangles. There were many steps in this hypothesis that failed to have confirmation and required leaps of faith that they were occurring. Evidence in support of the amyloid hypothesis came from the discovery of the genes involved in amyloid production: mutations in three genes formed the basis for the familial forms of AD; APP, presenilin 1 (PSEN1), and presenilin 2 (PSEN2); these three genes cause the rare familial form of AD, resulting in onset between 30 and 50 years of age. More common are the genes involved in the late onset forms of AD; the APOE genes that have three variants: E2, E3, and E4. Other genes have been shown to be involved in large genome-wide studies, but these have not proven to be as useful in understanding the pathophysiology as the major amyloid and APOE genes. A major blow to the amyloid hypothesis occurred with the publication of several studies of treatment with antibodies that successfully reduced amyloid in the brain as shown by PET studies, but failed to slow the cognitive decline. In addition, agents that blocked the secretases appeared to worsen the disease process (Anderson et al., 2017). As the amyloid hypothesis came under increased pressure, which was based on the failure of a large number of trials of agents that reduced amyloid without improving cognition, alternative theories of the etiology of Alzheimer's disease filled the void, which has led to an expansion of the disease pathways considered to be important (Hardy, 2017). This has produced a plethora of possible pathways that include most of the major disease producing processes; this large number of processes is more confusing than helpful as shown in a recent attempt to show all the possible contributions to the pathology (Musiek & Holtzman, 2015). The resulting void has opened the gates to alternative theories.

Alternative Alzheimer's disease hypotheses: blood–brain barrier dysfunction

In addition to deposition of amyloid in plaques around blood vessels and buildup of p-Tau in dying neurons, a number of other pathological processes take place with aging. The major changes related to increasing age are in blood vessels, while other proteins include α-synuclein associated with Parkinson's disease and a newly discovered protein, TDP-43. Examination of autopsies in the ADRC brain banks identified a high percentage of both plaques, tangles, and vascular disease (Toledo et al., 2013). Patients in a brain bank from the community found a similar high percentage (Arvanitakis et al., 2016; Schneider et al., 2007). Besides the autopsy data, there is MRI evidence of vascular disease; cerebral blood flow is reduced, diffusion tensor imaging shows ischemic damage to the white matter. There are also metabolic studies with PET, using positron-labeled FDG-PET. FDG is only partially metabolized and the PET signal indicates where glucose metabolism was occurring, making it a biomarker that indicates where there is normal glucose metabolism.

The buildup of amyloid plaques is a slow process that occurs over 20 years. Amyloid PET scans show that about 20% of normal individuals have amyloid deposits. Similarly, some healthy people have abnormal amyloid in

CSF. This creates problems when PET and CSF amyloid are used to select patients for studies. An important early study of a group of sisters in one religious order was the first to show the impact of vascular disease on cognitive function on those individuals with amyloid detected at autopsy. This study, referred to as the Nun Study or Religious Order Study, recruited a large number of participants who agreed to undergo yearly evaluations and to donate their brains for neuropathological study. The ones that had the worst cognitive function were those with both amyloid and vascular damage. This was the first indication that dual pathology was important to study and that vascular disease may play a major role in AD (Snowdon et al., 1997).

A number of mechanisms can lead to blood vessel damage in AD, including cerebral amyloid angiopathy (CAA), arteriolosclerosis/lipohyalinosis (AS/LH), and hereditary diseases. CAA begins in the cortex and progresses to deeper structures; AS/LH initially affects the basal ganglia, evolving into the deep white matter and thalamus, and eventually including the brainstem (Thal et al., 2003). Arteriolosclerosis is a prominent finding in small vessel disease; it is associated with leakage of albumin into the CSF. In a study of BBB function in 85-year-olds, increased levels of albumin in the CSF were commonly found in both AD and vascular dementia, indicating the prevalence of vascular disease in both these diseases in the elderly (Skoog et al., 1998).

Toledo and colleagues compared the prevalence of cerebrovascular disease, vascular pathology, and vascular risk factors in a wide range of neurodegenerative diseases and correlated them with dementia severity. They studied 5715 cases from the National Alzheimer's Coordinating Center database with a single neurodegenerative disease diagnosis (AD, frontotemporal lobar degeneration due to tau, and TAR DNA-binding protein 43 immunoreactive deposits, α-synucleinopathies, hippocampal sclerosis and prion disease). They determined the presence of cerebrovascular disease as defined by neuropathology and vascular pathology. They found that AD had the highest prevalence of vascular pathology and cerebrovascular disease than the other diagnoses. They concluded that concurrent cerebrovascular disease is a common neuropathological finding in aged subjects with dementia; is more common in AD than in other neurodegenerative disorders, especially in younger subjects; and lowers the threshold for dementia due to AD (Fig. 3.4) (Toledo et al., 2013).

The BBB is adversely affected in dementia. While this is particularly important in vascular dementia, it has been proposed as a major causative factor in AD. The evidence comes mainly from a study of the BBB permeability in the hippocampus of patients with MCI and AD. Blood vessels in the hippocampus are vulnerable to injury when the cerebral blood flow is compromised and the hippocampus is hypoperfused. One possible reason for this increased vulnerability is that the arteries supplying the hippocampus are end-arteries with the source of blood coming from the posterior cerebral and the anterior choroidal arteries. These posterior circulation arteries have to

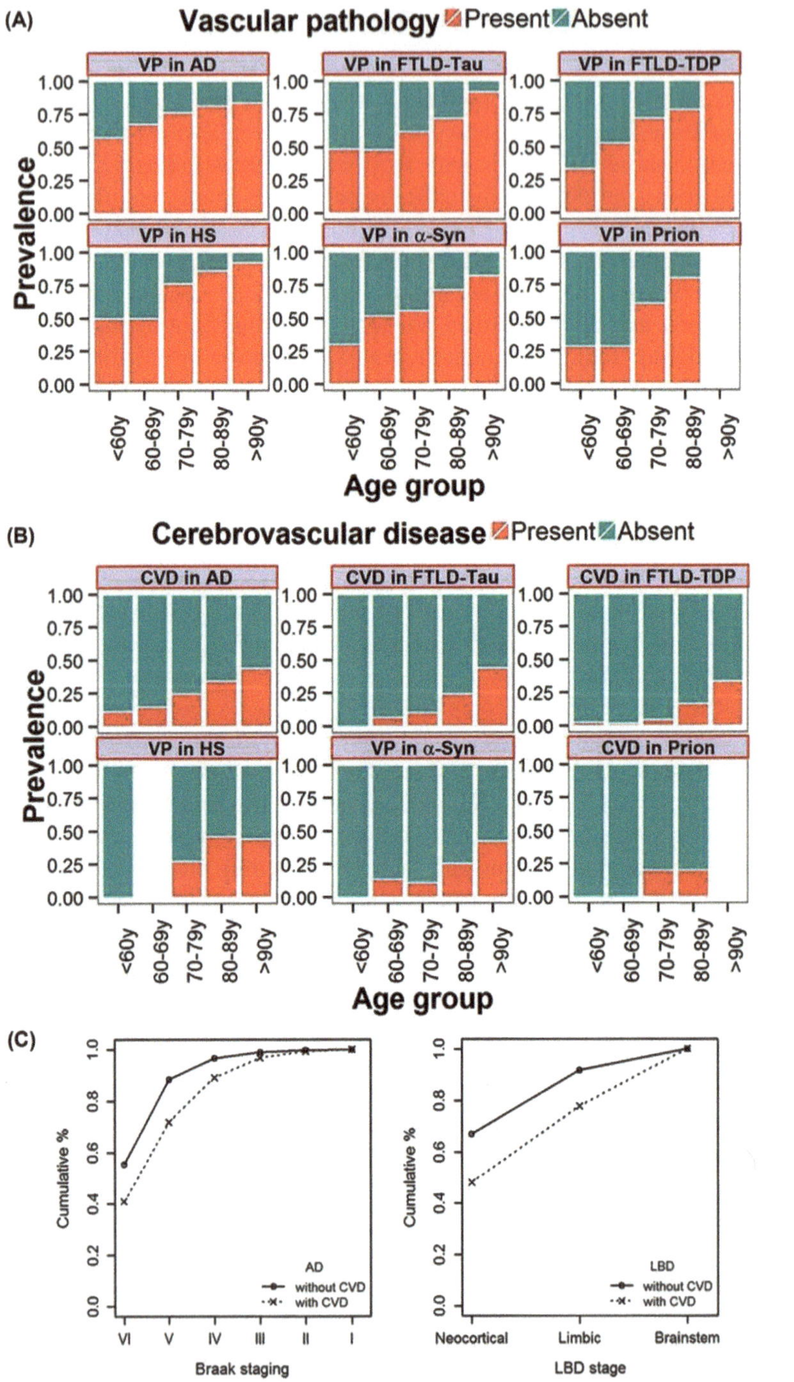

FIGURE 3.4 Prevalence of vascular pathology (A) and cerebrovascular disease (B) in the different neuropathologically diagnosed groups, and Braak stage (C left) and LBD stage (C right) stratified by the presence of cerebrovascular disease. *AD*, Alzheimer's disease; *CVD*, cerebrovascular disease; *HS*, hippocampal sclerosis; *LBD*, Lewy body disease stage; *y*, years.

travel a long distance as they supply blood to the hippocampus. Hippocampal vasculature at 7 T MRI and the autopsy image of the same region indicate the long path the vessels make, causing a reduction in blood flow, leading to hypoxia (Spallazzi et al., 2019). Medial temporal lobe dependent cognitive functions are highly vulnerable to hypoxia in the hippocampal region.

The hippocampal region is a site of BBB disruption. Aging, MCI, and AD increase the permeability of the vessels in the hippocampus. With high resolution MRI, the various regions of the hippocampus can be visualized during dynamic contrast-enhanced MRI (DCEMRI), and the permeability calculated for the individual regions. Using high spatial and temporal resolution MRI focused on the hippocampus, Zlokovic and colleagues found an age-dependent BBB breakdown in the hippocampus and CA1 and dentate gyrus subdivisions; this correlated with MCI. CSF studies showed elevated levels of the pericyte cell biomarker, soluble platelet-derived growth factor receptor-β, and the disruption of the BBB increased with age and diagnosis being more prominent in MCI (Fig. 3.5) (Montagne et al., 2015).

Opening of the BBB outside of the hippocampus has been demonstrated in patients with AD. Normally, the tight junctions in blood vessels prevent most substances from leaving the blood and entering the brain. Abnormal permeability of the BBB allows proteins and water from the blood to enter, interfering with brain function. In a study of BBB function in early AD, using DCEMRI, the investigators found that leakage was higher in patients than in controls; the increase in BBB damage was located in the gray matter, but was also found in normal appearing white matter (van de Haar et al., 2016). The multiplicity of factors that could potentially be involved in producing the symptoms in AD and the lack of long-term data make it difficult to determine the earliest factors influencing the symptoms. Accumulation of amyloid begins 20 years before symptoms develop and before the phosphylated tau appears. It is unlikely that the BBB is damaged at those early stages. Atrophy is also a late stage event and may be related to the toxic effects of Aβ and phosphoTau. As cells die, needs are reduced and cerebral blood flow decreases, resulting in a hypoxic environment. These events likely occurred prior to the opening of the BBB, which aggravates cell death and hypoxia. Since damage to the blood vessel requires an initiating event, another pathological process is needed. One possibility is that the aging of the blood vessels, which is known to accelerate loss of vascular integrity, could be involved; the presence of vascular risk factors, such as hypertension and diabetes, could be involved. The combination of toxic AD proteins, aging blood vessels, and vascular risk, which is occurring in patients with dual pathologies or mixed dementia, may be important in augmenting the vessel damage: in that case, those patients with both processes would have the most severe form of cognitive loss; it would also make it more problematic in determining the primary events.

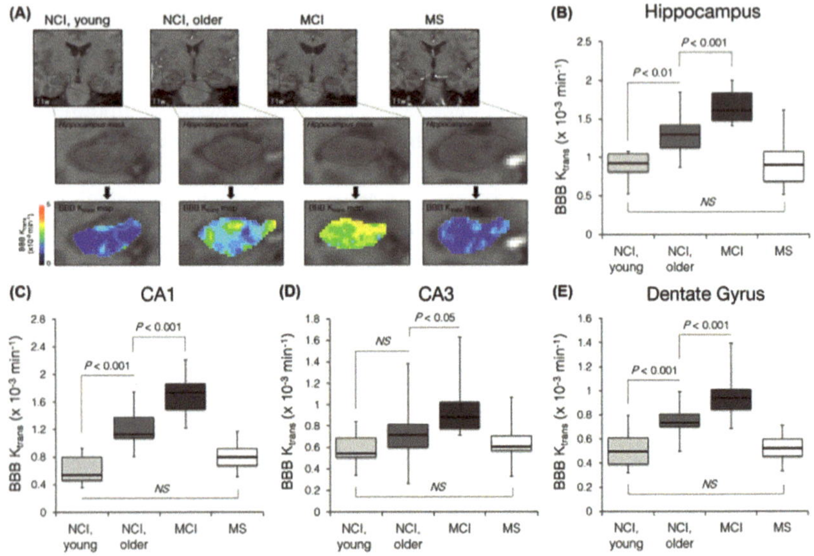

FIGURE 3.5 **BBB breakdown in the hippocampus.** BBB breakdown in the hippocampus during normal aging and aging associated with MCI (A) representative K_{trans} maps within the hippocampus in young and older individuals with NCI and MCI. MS, a multiple sclerosis case with no cognitive impairment. (B–E) A progressive significant increase in the BBB permeability constant K_{trans} in the older NCI group and MCI compared to the older NCI group in the entire hippocampus, CA1 region, and dentate gyrus. The MS group was compared with the age-matched young NCI group. Boxplots represent the median (*dark horizontal line*), with the box representing the 25th and 75th percentiles, the whiskers the 5th and 95th percentiles. p, significance by analysis of variance (ANOVA) followed by Tukey's post hoc tests. NCI, young ($n = 6$, ages 23–47, both genders); NCI, older ($n = 18$, ages 55–91, both genders); MCI ($n = 20$, ages 55–85, both genders); Ms ($n = 19$, ages 26–53, both genders).

Positron emission tomography biomarkers

There is a growing number of PET-based ligands that can be used to identify the various forms of dementia. They range from specific radioligands that bind to amyloid, tau and other proteins to nonspecific ones that show regions of abnormal metabolic activity such as deoxyglucose. The best studied are tracers for amyloid that indicate the regions of the brain where the protein in accumulating. More recently, tracers to identify tau have been developed; these are more relevant to disease progression and cell death. Many studies use combinations of these ligands, since none have been shown to be optimal.

Detection of protein pathology in the brain with molecular imaging is considered the "gold standard" by which the accuracy of blood-based biomarkers is tested. Therefore, to understand how Aβ deposition, tau deposition, and glucose metabolism are visualized with molecular imaging techniques and how these measures differ between different neurodegenerative disorders is important for fluid biomarker research (Fig. 3.2). In Fig. 3.2, the top row shows axial slices of ^{11}C-Pittsburgh compound B (^{11}C-PiB) scans, which reflect neuritic Aβ plaque density, from four patients. Higher levels of tracer retention are indicated by a higher standardized uptake value ratio (SUVR). The scan of the individual with AD shows considerable tracer retention throughout the cortex, especially when compared with the nonspecific tracer retention in the white matter, and is considered "Aβ-positive." The ^{11}C-PiB–PET scans in the individuals with behavioral variant FTD (bvFTD), corticobasal syndrome (CBS) or progressive supranuclear palsy (PSP) only show nonspecific tracer retention, which is localized to the white matter, and are therefore considered "Aβ-negative." The second row shows PET images acquired using the tracer ^{18}F-flortaucipir (^{18}F-FTP), which binds to intracellular aggregates of abnormal p-Tau. Tracer binding in the individual with AD is high in temporoparietal areas, including the posterior cingulate and precuneus, as well as the dorsal prefrontal cortex. In the individuals with bvFTD, CBS or PSP, the arrowheads highlight areas of mild–moderate tracer binding. Asterisks indicate brain regions of nonspecific tracer retention ("off-target" binding), which overlap with known patterns of off-target binding in healthy individuals, including binding in the basal ganglia, midbrain regions and cerebellum. The third row shows PET images acquired using the tracer ^{18}F-FDG, the retention of which is a marker for glucose uptake by the tissue. Across all individuals, regions of low ^{18}F-FDG retention overlap with regions of high ^{18}F-FTP retention. The bottom row shows structural T1-weighted MRI scans that correspond to the slices shown for the ^{18}F-FTP and ^{18}F-FDG scans.

The PiB–PET scans were acquired 50–70 minutes after tracer injection, and SUVRs were created using a cerebellar reference region; the ^{18}F-FTP scans were acquired at 80–100 minutes after tracer injection and SUVR created using an inferior cerebellar reference region; the ^{18}F-FDG scans were acquired at 30–60 minutes after tracer injection and SUVR created using the pons as the reference region.

PET provides exquisite anatomical detail with some overlap between different forms of dementia. For many clinical trials, it is used as the major biomarker to select patients and in some, it can be used to indicate progression or regression of the tracer. The drawbacks are the high cost, unavailability of PET scanners, the need to locally produce or ship the tracers, and the exposure to radioactivity. Except for FDG, the other tracers are mainly used for research purposes. With the development of methods to sample proteins in the blood, which are likely to replace PET scanning except for determining the presence of the tracers to select patients and regression of the proteins to show efficacy of the treatments.

Blood-based biomarkers

Biomarkers are improving the ability to diagnose different forms of dementia. For AD, the biomarkers from CSF and PET have been used for many years and much has been learned about their ability to both identify those with the disease and to predict those that are destined to progress. Vascular dementia has been a more difficult problem with no universally accepted biomarkers to use in diagnosis. While PET studies of amyloid and tau have come the closest to the actual disease, since they measure levels of these two important proteins in the brain. In addition, they have been used to verify the accuracy of the same biomarkers in the CSF. The main problem has been the overlap with positive results for the tracers in normal subjects; it is possible to have major accumulations of amyloid and tau without causing cognitive loss and vice versa. Autopsy studies in a group over 90 years of age made this clear in that the level of dementia was not directly correlated with the level of these proteins in the brain, suggesting that other factors were driving the accumulation of the abnormal proteins. This makes it more important to use a number of biomarkers in conjunction with the clinical findings rather than rely on one or another of the biomarkers alone.

A new group of biomarkers is emerging as potentially very important; these are the plasma-based biomarkers that are measured with a number of ultrasensitive assays. The SIMOA developed by Quanterix Corp. is currently the most promising. Many studies done with this and other instruments have correlated the levels of species of amyloid and p-Tau in the blood with those measured in CSF and PET; they have shown a high level of correlation, suggesting that these assay methods will soon replace the invasive and expensive older methods. This could allow for large scale screening of populations to identify those at high risk of developing dementia, which in turn could be used to start treatments at an earlier stage of the illness. These ultrasensitive instruments, in addition to measuring amyloid and tau, can also detect other molecules that could help unravel the underlying pathophysiology. The most promising of these other assays are neurofilament light (NfL), which is a nonspecific indicator of axonal damage in several neurological disorders, and glial fibrillary acidic protein (GFAP), which is released by activated astrocytes. Blood P-tau_{181} could be used as a simple, accessible, and scalable test for screening and diagnosis of AD (Karikari et al., 2020). Another analysis method is quantitative mass spectrometry, which is a sensitive method to detect in CSF two p-Tau isoforms, $Ptau_{181}$ and $Ptau_{217}$; the 217 isoform is more accurate than the 181 form in identifying AD (Barthélemy, Horie, et al., 2020).

A recent study showed that tau phosphorylated at threonine 231 ($Ptau_{231}$) is an excellent blood biomarker for dementia that can be measured with the SIMOA ELISA assay. It has been validated in four independent cohorts ($n = 588$) in different settings; plasma $Ptau_{231}$ was able to identify patients with AD and to distinguish AD patients from patients with non-AD neurodegenerative disorders, including Aβ negative MCI patients. More importantly, in a neuropathology cohort, plasma $Ptau_{231}$ in samples taken on average 4.2 years prior to postmortem, very accurately identified AD neuropathology in comparison to non-AD neurodegenerative disorders. Plasma $Ptau_{231}$ was highly correlated with CSF $Ptau_{231}$. Thus, plasma $Ptau_{231}$ assay identifies the clinical stages of AD and neuropathology equally well as plasma $Ptau_{181}$, but increases earlier when subtle Aβ deposition has begun, prior to the threshold for Aβ PET positivity has been attained, and also in response to early brain tau deposition (Ashton et al., 2021).

NfL chain is another molecule that can be detected in the blood by ultrasensitive assay methods; it is the smallest of the neurofilament triplet proteins that are the structural components of axons. During healthy aging, NfL is released from axons and enters the extracellular space, which results in increased CSF NfL concentrations. When axons are injured, NfL release is accelerated, which can be caused by neurodegeneration, inflammation, trauma, or stroke. Several studies have shown that CSF NfL concentrations are highest in brain disorders with subcortical pathology, such as vascular dementia and normal pressure hydrocephalus. They are particularly elevated in CSF in individuals with frontotemporal dementia than in individuals with AD. In addition, CSF NfL concentrations are substantially higher in individuals with Creutzfeldt–Jakob disease than in healthy individuals, which correlates with the changes in CSF tTau concentrations observed in this disease. SIMOA immunoassays can detect NfL in the plasma, making it another of the blood-based biomarkers for dementia (Mattsson et al., 2019).

Selection of biomarkers: cerebrospinal fluid versus blood

The availability of an array of assay methods with various highly accurate biomarkers promises to have a major impact on diagnosis of dementia and evaluation of clinical trials. While CSF remains the most accurate because of the continuity of CSF in brain tissue, the accuracy of the new blood-based biomarkers is reaching that of CSF. With more time for testing in different populations, it is likely that blood will become the fluid of choice. Eliminating the need for lumbar puncture, which is an invasive test, and PET scanning that is both costly and of limited availability, is highly desirable. There are many biomarkers that remain measurable only in the CSF that will be needed for research studies, but more than likely the routine clinical work will be done with blood-based assays. This will allow large scale screening of populations to identify those individuals at higher risk; this group will be recruited into clinical trials of agents that can be tested at earlier stages of the disease.

Currently, the two major AD proteins, amyloid and tau, are the only ones available for blood-based assay. However, assays now being tested for a wide variety of disease processes will reach clinical usefulness. Other assays under development include GFAP, which is a biomarker for astrocyte injury. A number of inflammatory biomarkers, including proteases and cytokines, will be available for detection of ongoing inflammation. Eventually, genetic biomarkers in the blood will achieve clinical utility. This will create another level of problems that will result in the greater use of machine learning approaches with artificial intelligence applications similar to those being developed for precision medicine in oncology.

References

Anderson, R. M., Hadjichrysanthou, C., Evans, S., & Wong, M. M. (2017). Why do so many clinical trials of therapies for Alzheimer's disease fail? *Lancet (London, England)*, *390*(10110), 2327–2329. Available from https://doi.org/10.1016/S0140-6736(17)32399-1.

Arvanitakis, Z., Capuano, A. W., Leurgans, S. E., Bennett, D. A., & Schneider, J. A. (2016). Relation of cerebral vessel disease to Alzheimer's disease dementia and cognitive function in elderly people: A cross-sectional study. *The Lancet Neurology*, *15*(9), 934–943. Available from https://doi.org/10.1016/S1474-4422(16)30029-1.

Ashton, N. J., Hye, A., Rajkumar, A. P., Leuzy, A., Snowden, S., Suárez-Calvet, M., Karikari, T. K., Schöll, M., La Joie, R., Rabinovici, G. D., Höglund, K., Ballard, C., Hortobágyi, T., Svenningsson, P., Blennow, K., Zetterberg, H., & Aarsland, D. (2020). An update on blood-based biomarkers for non-Alzheimer neurodegenerative disorders. *Nature Reviews Neurology*, *16*(5), 265–284. Available from https://doi.org/10.1038/s41582-020-0348-0.

Ashton, N. J., Pascoal, T. A., Karikari, T. K., Benedet, A. L., Lantero-Rodriguez, J., Brinkmalm, G., Snellman, A., Schöll, M., Troakes, C., Hye, A., Gauthier, S., Vanmechelen, E., Zetterberg, H., Rosa-Neto, P., & Blennow, K. (2021). Plasma p-tau231: A new biomarker for incipient Alzheimer's disease pathology. *Acta Neuropathologica*, *141*(1–15), 1–16. Available from https://doi.org/10.1007/s00401-021-02275-6.

Barthélemy, N. R., Horie, K., Sato, C., & Bateman, R. J. (2020). Blood plasma phosphorylated-tau isoforms track CNS change in Alzheimer's disease. *The Journal of Experimental Medicine*, *217*(11). Available from https://doi.org/10.1084/jem.20200861.

Barthélemy, N. R., Li, Y., Joseph-Mathurin, N., Gordon, B. A., Hassenstab, J., Benzinger, T. L. S., Buckles, V., Fagan, A. M., Perrin, R. J., Goate, A. M., Morris, J. C., Karch, C. M., Xiong, C., Allegri, R., Mendez, P. C., Berman, S. B., Ikeuchi, T., Mori, H., Shimada, H., & McDade, E. (2020). A soluble phosphorylated tau signature links tau, amyloid and the evolution of stages of dominantly inherited Alzheimer's disease. *Nature Medicine*, *26*(3), 398–407. Available from https://doi.org/10.1038/s41591-020-0781-z.

Blennow, K., Hampel, H., Weiner, M., & Zetterberg, H. (2010). Cerebrospinal fluid and plasma biomarkers in Alzheimer disease. *Nature Reviews Neurology*, *6*(3), 131–144. Available from https://doi.org/10.1038/nrneurol.2010.4.

Glenner, G. G., & Wong, C. W. (1984). Alzheimer's disease: Initial report of the purification and characterization of a novel cerebrovascular amyloid protein. *Biochemical and Biophysical Research Communications*, *120*(3), 885–890.

Goedert, M. (2009). Oskar Fischer and the study of dementia. *Brain : A Journal of Neurology*, *132*(Pt 4), 1102–1111. Available from https://doi.org/10.1093/brain/awn256.

Greenberg, S. M., Bacskai, B. J., Hernandez-Guillamon, M., Pruzin, J., Sperling, R., & van Veluw, S. J. (2020). Cerebral amyloid angiopathy and Alzheimer disease – One peptide, two pathways. *Nature Reviews Neurology*, *16*(1), 30–42. Available from https://doi.org/10.1038/s41582-019-0281-2.

Hardy, J. (2017). The discovery of Alzheimer-causing mutations in the APP gene and the formulation of the "amyloid cascade hypothesis.". *The FEBS Journal*, *284*(7), 1040–1044. Available from https://doi.org/10.1111/febs.14004.

Hardy, J., & Selkoe, D. J. (2002). The amyloid hypothesis of Alzheimer's disease: Progress and problems on the road to therapeutics. *Science (New York, N.Y.)*, *297*(5580), 353–356.

Jack, C. R., Bennett, D. A., Blennow, K., Carrillo, M. C., Dunn, B., Haeberlein, S. B., Holtzman, D. M., Jagust, W., Jessen, F., Karlawish, J., Liu, E., Molinuevo, J. L., Montine, T., Phelps, C., Rankin, K. P., Rowe, C. C., Scheltens, P., Siemers, E., Snyder, H. M., & Sperling, R. (2018). NIA-AA research framework: Toward a biological definition of Alzheimer's disease. *Alzheimer's & Dementia : The Journal of the Alzheimer's Association*, *14*(4), 535–562. Available from https://doi.org/10.1016/j.jalz.2018.02.018.

Jack, C. R., & Holtzman, D. M. (2013). Biomarker modeling of Alzheimer's disease. *Neuron*, *80*(6), 1347–1358. Available from https://doi.org/10.1016/j.neuron.2013.12.003.

Jarrett, J. T., Berger, E. P., & Lansbury, P. T. (1993). The carboxy terminus of the beta amyloid protein is critical for the seeding of amyloid formation: Implications for the pathogenesis of Alzheimer's disease. *Biochemistry*, *32*(18), 4693–4697.

Karanth, S., Nelson, P. T., Katsumata, Y., Kryscio, R. J., Schmitt, F. A., Fardo, D. W., Cykowski, M. D., Jicha, G. A., Van Eldik, L. J., & Abner, E. L. (2020). Prevalence and clinical phenotype of quadruple misfolded proteins in older adults. *JAMA Neurology*, *77*(10), 1299–1307. Available from https://doi.org/10.1001/jamaneurol.2020.1741.

Karikari, T. K., Pascoal, T. A., Ashton, N. J., Janelidze, S., Benedet, A. L., Rodriguez, J. L., Chamoun, M., Savard, M., Kang, M. S., Therriault, J., Schöll, M., Massarweh, G., Soucy, J.-P., Höglund, K., Brinkmalm, G., Mattsson, N., Palmqvist, S., Gauthier, S., Stomrud, E., & Blennow, K. (2020). Blood phosphorylated tau 181 as a biomarker for Alzheimer's disease: A diagnostic performance and prediction modelling study using data from four prospective cohorts. *The Lancet Neurology*, *19*(5), 422–433. Available from https://doi.org/10.1016/S1474-4422(20)30071-5.

Lee, J.-M., Yin, K.-J., Hsin, I., Chen, S., Fryer, J. D., Holtzman, D. M., Hsu, C. Y., & Xu, J. (2003). Matrix metalloproteinase-9 and spontaneous hemorrhage in an animal model of cerebral amyloid angiopathy. *Annals of Neurology*, *54*(3), 379–382.

Mattsson, N., Cullen, N. C., Andreasson, U., Zetterberg, H., & Blennow, K. (2019). Association between longitudinal plasma neurofilament light and neurodegeneration in patients with Alzheimer disease. *JAMA Neurology*, *76*(7), 791–799. Available from https://doi.org/10.1001/jamaneurol.2019.0765.

McKhann, G., Drachman, D., Folstein, M., Katzman, R., Price, D., & Stadlan, E. M. (1984). Clinical diagnosis of Alzheimer's disease: Report of the NINCDS-ADRDA Work Group* under the auspices of Department of Health and Human Services Task Force on Alzheimer's Disease. *Neurology*, *34*(7), 939–944. Available from https://doi.org/10.1212/WNL.34.7.939.

Montagne, A., Barnes, S. R., Sweeney, M. D., Halliday, M. R., Sagare, A. P., Zhao, Z., Toga, A. W., Jacobs, R. E., Liu, C. Y., Amezcua, L., Harrington, M. G., Chui, H. C., Law, M., & Zlokovic, B. V. (2015). Blood-brain barrier breakdown in the aging human hippocampus. *Neuron*, *85*(2), 296–302. Available from https://doi.org/10.1016/j.neuron.2014.12.032.

Musiek, E. S., & Holtzman, D. M. (2015). Three dimensions of the amyloid hypothesis: Time, space and "wingmen.". *Nature Neuroscience*, *18*(6), 800–806. Available from https://doi.org/10.1038/nn.4018.

Petersen, R. C. (2016). Mild cognitive impairment. *Continuum (Minneapolis, MN)*, *22*(2 Dementia), 404–418. Available from https://doi.org/10.1212/CON.0000000000000313.

Raz, L., Yang, Y., Thompson, J., Hobson, S., Pesko, J., Mobashery, S., Chang, M., & Rosenberg, G. (2018). MMP-9 inhibitors impair learning in spontaneously hypertensive rats. *PLoS One*, *13*(12), e0208357. Available from https://doi.org/10.1371/journal.pone.0208357.

Schneider, J. A., Arvanitakis, Z., Bang, W., & Bennett, D. A. (2007). Mixed brain pathologies account for most dementia cases in community-dwelling older persons. *Neurology*, *69*(24), 2197–2204. Available from https://doi.org/10.1212/01.wnl.0000271090.28148.24.

Selkoe, D. J. (2019). Alzheimer disease and aducanumab: Adjusting our approach. *Nature Reviews Neurology*, *15*(7), 365–366. Available from https://doi.org/10.1038/s41582-019-0205-1.

Skoog, I., Wallin, A., Fredman, P., Hesse, C., Aevarsson, O., Karlsson, I., Gottfries, C. G., & Blennow, K. (1998). A population study on blood-brain barrier function in 85-year-olds: Relation to Alzheimer's disease and vascular dementia. *Neurology*, *50*(4), 966–971.

Snowdon, D. A., Greiner, L. H., Mortimer, J. A., Riley, K. P., Greiner, P. A., & Markesbery, W. R. (1997). Brain infarction and the clinical expression of Alzheimer disease. The Nun Study. *JAMA: The Journal of the American Medical Association*, *277*(10), 813–817. Available from https://doi.org/10.1001/jama.277.10.813.

Spallazzi, M., Dobisch, L., Becke, A., Berron, D., Stucht, D., Oeltze-Jafra, S., Caffarra, P., Speck, O., & Düzel, E. (2019). Hippocampal vascularization patterns: A high-resolution 7 Tesla time-of-flight magnetic resonance angiography study. *NeuroImage Clinical*, *21*, 101609. Available from https://doi.org/10.1016/j.nicl.2018.11.019.

Terry, R. D., & Katzman, R. (1983). Senile dementia of the Alzheimer type. *Annals of Neurology*, *14*(5), 497–506.

Thal, D. R., Ghebremedhin, E., Orantes, M., & Wiestler, O. D. (2003). Vascular pathology in Alzheimer disease: Correlation of cerebral amyloid angiopathy and arteriosclerosis/lipohyalinosis with cognitive decline. *Journal of Neuropathology and Experimental Neurology*, *62*(12), 1287–1301.

Toledo, J. B., Arnold, S. E., Raible, K., Brettschneider, J., Xie, S. X., Grossman, M., Monsell, S. E., Kukull, W. A., & Trojanowski, J. Q. (2013). Contribution of cerebrovascular disease in autopsy confirmed neurodegenerative disease cases in the National Alzheimer's Coordinating Centre. *Brain*, *136*(9), 2697–2706. Available from https://doi.org/10.1093/brain/awt188.

van de Haar, H. J., Burgmans, S., Jansen, J. F. A., van Osch, M. J. P., van Buchem, M. A., Muller, M., Hofman, P. A. M., Verhey, F. R. J., & Backes, W. H. (2016). Blood-brain barrier leakage in patients with early Alzheimer disease. *Radiology*, *281*(2), 527–535.

van Harten, A. C., Mielke, M. M., Swenson-Dravis, D. M., Hagen, C. E., Edwards, K. K., Roberts, R. O., Geda, Y. E., Knopman, D. S., & Petersen, R. C. (2018). Subjective cognitive decline and risk of MCI: The Mayo Clinic Study of Aging. *Neurology*, *91*(4), e300–e312. Available from https://doi.org/10.1212/WNL.0000000000005863.

Weller, R. O., Subash, M., Preston, S. D., Mazanti, I., & Carare, R. O. (2008). Perivascular drainage of amyloid-beta peptides from the brain and its failure in cerebral amyloid angiopathy and Alzheimer's disease. *Brain Pathology (Zurich, Switzerland)*, *18*(2), 253–266. Available from https://doi.org/10.1111/j.1750-3639.2008.00133.x.

Chapter 4

Classification of subtypes of vascular dementia

Definitions

The definition of vascular dementia has evolved since its initial conception as an accumulation of tissue damage following strokes (Blessed et al., 1968). As a consequence of the arguments over the correct definition, progress in vascular dementia research has been impeded by lack of agreement as to the definition and the mechanisms. However, more recently there has been progress in clarifying the definition and delineating different types. As opposed to the terminological controversy about vascular dementia, research into Alzheimer's disease (AD) has benefited from both early agreement on the definition of the disease and from a natural history that showed a predictable decline. Such benefits were not available in dementia related to vascular disease. In fact, several early concepts were misleading. In addition, the development of neuroimaging methods beginning in the 1970s resulted in major modifications of disease definitions. The earliest attempts to define vascular dementia were based mainly on the presence of strokes; it was thought that the larger the amount of brain tissue damaged by the stroke, the more likely for dementia to be present (Blessed et al., 1968). This concept ignored a large group of patients with vascular disease related to dementia due to disease of the small vessels that did not have major strokes. Prior to neuroimaging, damage to the white matter that was subsequently to be revealed by CT and MRI had been described in pathological studies beginning around the turn of the 19th century, first with Alzheimer's report of the abnormalities in the cortex and the description by Otto Binswanger of vascular demyelination related to arteriolosclerosis. However, the lack of interest in the neuropathological study of diseases of old age has limited progress to a few further reports.

The first breakthrough came with the application of isotopic methods to measure cerebral blood flow (CBF), which led to the concept that dementia was related to reduced CBF with hypoxic hypoperfusion: these studies showed that patients with abnormal CBF were at a higher risk of developing dementia. In one of the earliest studies with the newly developed method to measure CBF, 24 patients of comparable age, blood pressure, and degree of

Neuroinflammation in Vascular Dementia. DOI: https://doi.org/10.1016/B978-0-12-823455-6.00010-9

dementia were classified by an "Ischemic Score" based on clinical features into "multi-infarct dementia (MID)" and "primary degenerative" dementia; regional CBF was measured by the intracarotid injection of xenon^{133}. Both groups showed a decreased proportion of rapidly clearing brain tissue (largely gray matter). Studies of blood flow showed that CBF was normal in the primary degenerative disease group, but reduced in those with multiple strokes. This suggests that the blood flow was adequate for the metabolic needs of the brain in patients with primary degenerative dementia, but inadequate for those with MID. Furthermore, there was no correlation between the degree of dementia and the level of CBF in the primary degenerative group, but an inverse relationship was found in the multi-infarct group. Reactivity to changes in arterial carbon dioxide pressure was normal in both groups. The investigators concluded that the typical, insidious, slowly progressive dementia of old age was unlikely to be due to atherosclerosis (Hachinski et al., 1975); they found no relationship between parenchymal degenerative changes associated with AD and large vessel arterial disease. The term "cerebral atherosclerosis" became obsolete since it was misleading. It was replaced with the concept of multiple small or large cerebral infarcts (MID). This new information rekindled interest in the precise diagnosis of all cases of dementia, allowing for more accurate diagnoses and more rational management of the significant number of cases in which dementia was due to vascular disease, which is potentially a treatable condition (Hachinski et al., 1974).

The Hachinski Ischemic Score was developed to indicate those patients with a vascular etiology to the cognitive decline. For many years, the concept that vascular dementia was related to multiple strokes remained prevalent until the use of neuroimaging was dramatically increased; many people were found to have a progressive dementia unrelated to strokes, but rather to atrophy of the gray matter and/or injury to the white matter that could be seen on CT and MRI. With the direct proof of abnormal white matter in a number of cases of dementia, the mechanisms involved in white matter injury became a topic of intense research.

The first indication of the importance of changes in the white matter came from CT scans that showed attenuation of the white matter in many elderly individuals; white matter injury resulted in an attenuation of the X-ray signal which suggested increased water or decreased tissue, either of which would allow increased penetration of the X-ray signals. These early CT scans were consistent with Binswanger's pathological studies, and when the white matter injury appeared to be extensive, the patients were said to have Binswanger's disease. A problem was noted immediately that there were patients with CT evidence of white matter injury who were cognitively normal, making it improper to use a radiological finding as the basis for a disease diagnosis. The British medical journal, *Lancet*, wrote in an editorial that there was an apparent "epidemic" of Binswanger's disease. They advised against using a radiological finding as the basis for a diagnosis of dementia. After the use of MRI became widespread in clinical practice, the marked increase in sensitivity of MRI compared to CT, rather than reducing the overdiagnosis problem, was made worse. Overinterpretation of the MRI scans led to normal individuals being told they had Binswanger's disease based on the radiological findings (Kinkel et al., 1985).

An MRI pulse sequence entered common usage that was able to detect water in the tissue while removing the signal from water in the ventricles; this was fluid attenuated inversion discovery (FLAIR), and a large number of abnormal neurological disorders were found to have white matter hyperintensities (WMHs) since the water in the tissue was seen as a hyperintense signal. Although this was a non-specific finding that was also seen in both normal healthy subjects and those with vascular risk factors, primarily hypertension, it was not a sign of dementia. Many patients were told they had "ischemic injury" based on a radiological finding that was not necessarily related to an abnormal neurological examination or actual symptoms. Hachinski used a term derived from two Latin words to describe the changes in the white matter that were seen radiologically: he coined the term leukoaraiosis derived from the Latin words, *leuko* meaning white and *araiosis* meaning pale (Hachinski et al., 1987). The original intent was that this would be a radiological descriptive term that lacked pathological significance. Although WMHs are associated with strokes and cognitive loss in large surveys, they are less helpful in the individual patients. Patients can have leukoaraiosis without any associated clinical abnormality. Hypertension, smoking, diabetes, hyperhomocysteinemia, and heart disease are all risk factors for leukoaraiosis; it can be an initial stage of Binswanger's disease, but this evolution does not always happen. White matter lesions on MRI are non-specific with many healthy, cognitively intact elderly showing large areas of WMHs; in one survey of healthy, community-dwelling subjects over the age 65 with neither vascular disease nor cognitive impairment, 30% had moderately large WMHs, and 7% had complete opacity of the white matter (Fig. 4.1) (Hunt et al., 1989).

Increased availability of CT and MRI scanning forced the clinicians to revise the definitions of vascular dementia in people with cognitive impairment, incorporating neuroimaging as an adjunct in the diagnosis. The definition of vascular dementia underwent a series of name changes, expanding the definition to include patients with extensive white matter injury, but without obvious strokes. This was incorporated into the subsequent definitions (Chui et al., 1992). This new category was added to the prior one of MID and was soon recognized to be related to "small vessel disease" (Pantoni, 2010). Eventually it was realized that the small vessel form of vascular dementia was the most prevalent form (Snyder et al., 2015). With the increased interest in dementia and the increase in the number of autopsies, pathologists have become aware that dementia is a disease process related to multiple etiologies with at least four known misfolded proteins identified pathologically in the

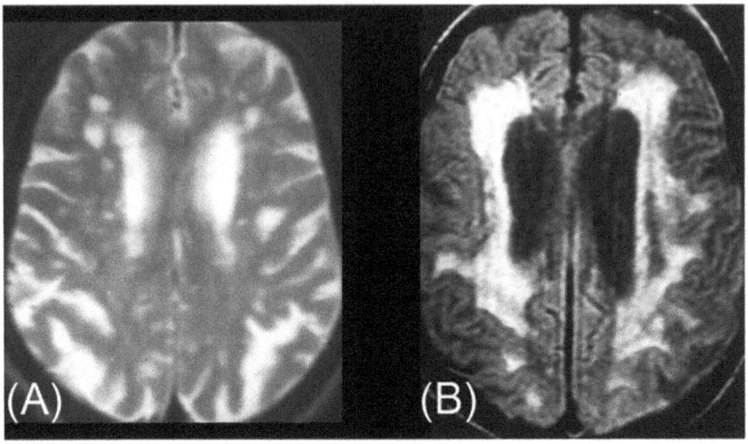

FIGURE 4.1 **FLAIR white matter hyperintensities in normal elderly.** A patient in an imaging study of normal aging in subjects over the age of 65. (A) FLAIR MRI of subject with moderate white matter hyperintensities, which were seen in 30%. (B) A patient over the age of 80 with severe white matter hyperintensities; severe white matter hyperintensities were found in 7% of these healthy controls with normal scores on neuropsychological testing. *FLAIR*, Fluid attenuated inversion discovery.

elderly (Karanth et al., 2020). These are related to misfolded proteins that have become hallmarks of different forms of dementia. The most common abnormal proteins are amyloid-β and phosphorylated tau that are the hallmarks of AD. Other abnormal proteins are α-synuclein, which is in the brains of patients with Parkinson's disease, and a recently described protein, transactive response DNA binding protein 43 kDa (TDP-43) that is increased in hippocampal sclerosis. Combinations of the abnormal proteins are commonly found particularly as patients increase in age (Karanth et al., 2020).

A new definition of vascular dementia was needed that included both multiple strokes and the absence of strokes with extensive injury to the white matter: modern neuroimaging led to the introduction of terms for non-infarct vascular changes, such as WMHs, subcortical lacunes and microbleeds, as contributors to cognitive decline. The term vascular cognitive impairment (VCI), an umbrella term for all the different types of vascular disease that could lead to dementia, was introduced (Hachinski et al., 2006), and VCI was related mainly to small vessel disease (Pantoni, 2010). Finally, the term VCI was expanded to include the final stages of actual dementia, and the currently used term—vascular cognitive impairment and dementia (VCID)—entered the nomenclature. This term is now the most popular since it includes the full spectrum of vascular causes of dementia, ranging from vascular mild cognitive impairment to dementia (Iadecola et al., 2019).

Controversy has surrounded the use of the term Binswanger's disease; although it indicated white matter disease associated with neurological and cognitive abnormalities, it is unclear if it should be limited to extensive white matter involvement or used more generally. In either case, it is a progressive form of white matter disease seen most often as a consequence of long-standing hypertension or other vascular diseases, including diabetes, hyperlipidemia, and hypoxia of sleep apnea; neuropathology indicates arteriolosclerosis of the penetrating arteries (Caplan & Schoene, 1978; Huisa & Rosenberg, 2014; Rosenberg et al., 1979). The white matter is demyelinated, appearing as an attenuated region on CT and a hyperintense area on MRI. Typical symptoms are apathy, depression, executive dysfunction, incontinence, and gait disturbance. On neurological examination there is often symmetric hyperreflexia, inability to walk on a straight line, and imbalance without cerebellar signs or muscle weakness. The MRI shows large areas of WMHs: over time there is gradual growth of the white matter lesions. The pattern of white matter increased signal corresponds with the regions that are vulnerable to reduced CBF.

Major types of vascular disease

Large vessel disease

Blood vessel size separates different types of vascular disease: large blood vessels are more commonly associated with strokes related to emboli from carotid artery due to atherosclerosis cholesterol plaques or from heart, mainly due to atrial fibrillation (AF) or valvular damage, while small vessel disease leads to white matter damage. Large strokes are also due to intracranial atherosclerosis; these strokes result in a sudden loss of cognitive ability, and in the case of multiple strokes, the drop in intellect can occur in a stepwise fashion. Included in the large vessel stroke group are those with a single stroke in a strategic brain region: these are usually in the thalamus and posterior corpus callosum and cause a change in memory function. Similarly, single strokes in the hippocampus can result in an amnestic syndrome: these are usually due to occlusion of the posterior cerebral artery, which can cause, along with amnesia, a field defect related to a damaged optic tract. Lacunar strokes in the basal ganglia produce motor and sensory loss, but rarely interfere with thinking unless they are multiple. Recovery from either large or small strokes is common with some residual dysfunction remaining. The pattern of cognitive loss after large strokes is one of loss of function with some recovery. When multiple strokes are involved, this appears as a stepwise loss of function with long intervals between events. Because of the erratic nature of the multiple stroke syndromes with progressive loss of intellect, sporadic stroke syndromes are difficult to characterize. Without a clear picture of the natural history, assessing treatments to improve cognition are not possible. On the contrary, small vessel disease, which often has a more predictable natural history, is optimal for treatment trials.

Thromboembolism from the carotid and the heart can cause transient ischemic attacks (TIAs) that are often multiple. Carotid stenosis due to atherosclerosis creates roughened surfaces that accumulate small clots, which can be released into the circulation, causing large and small strokes. Identification of a high-grade stenosis after one or more TIAs is an indication for aggressive medical treatment or surgical removal of the plaque. There are patients with high-grade stenosis discovered incidentally during screening procedures for atherosclerosis; some advocate prophylactic removal of the stenosis, but the optimal time for surgery remains controversial. Occasionally, a patient is identified with bilateral carotid artery stenosis that appears to be impacting cognition. Such patients can be seen to have an increase in WMHs in the centrum semiovale bilaterally; this is considered to be due to hypoperfusion affecting the watershed regions at the ends of the major arteries supplying both hemispheres. The role of surgery in these patients is an ongoing controversy (Buratti et al., 2014; Norling et al., 2019).

In a study of asymptomatic internal carotid artery stenosis, investigators used MRI-based perfusion and oxygenation-sensitive imaging. They found significant impairments of CBF, cerebral vascular reactivity (CVR), regional cerebral blood volume (rCBV), oxygen extraction capacity and capillary transit-time heterogeneity. Most pronounced changes of CBF, CVR and rCBV were detected in watershed areas. Complementary MRI-based perfusion and oxygenation parameters offer deeper perspectives on complex microvascular impairments in individual patients. Furthermore, knowledge of the physiological changes improves sensitivity of detection of hemodynamic impairments (Kaczmarz et al., 2021). In spite of many studies, the impact of carotid stenosis on cognition remains unresolved.

Emboli from the heart are another major source of strokes leading to cognitive decline. AF is the most common heart abnormality associated with dementia. A large study of elderly patients that separated patients into younger and older age ranges was done in Republic of Korea. However, the association between the two diseases is not well identified in different age subgroups. Younger participants with AF were more prone to dementia development than older participants with AF. AF was associated with an increased risk of both early- and late-onset dementia, independent of clinical stroke (Kim, Yang, Lip, et al., 2020). Anticoagulants prevent dementia in AF. Use of the newer direct oral anticoagulants as opposed to warfarin in AF-related dementia showed that they are beneficial in preventing dementia. Apixaban and rivaroxaban use were both associated with a significantly lower risk of dementia outcomes compared with warfarin (Lee et al., 2021). Catheter ablation for AF prolongs the duration of sinus rhythm, improving the quality of life. A study of the association of catheter ablation for AF with the occurrence of dementia showed that those treated with catheter ablation or medical therapy, ablation was associated with decreased dementia risk (Kim, Yang, Sung, et al., 2020).

A small group of patients with strokes have dementia related to a single event, particularly in those with infarcts of the posterior cerebral artery, which supplies the hippocampus and parts of the thalamus. The syndrome of amnesia with a field cut can be misdiagnosed when the amnesia is subtle and the field cut is not recognized (Benson et al., 1974). Occasionally small vessels supplying other posterior regions of the brain, such as the posterior corpus callosum can cause an amnestic type of dementia. Since these are often isolated vessels they are called single stroke dementia, and have a different pathophysiology and are generally sporadic, making them different from the usual small vessel disease.

Small vessel disease

Gradual changes occur in the blood vessels over long periods of time with aging; there is thickening of the vessel wall and narrowing of the lumen. Normal pulsations of the vessel are dampened, slowing the responses to changes in position and oxygen. These normal changes can be accelerated by hypertension, diabetes, hyperlipidemia, and hereditary diseases; many of these vascular diseases are modifiable by changes in lifestyle. Hypertension and diabetes are most commonly associated with small vessel disease with the greatest impact on the arterioles. Long-standing hypertension slowly remodels the blood vessels, which is often asymptomatic until the vessel is damaged beyond repair. At this late stage, the narrowing of the vessel lumen limits the blood flow to the brain, while the lack of responsiveness of the stiffened vessel further hampers delivery of blood to the brain. Gradual fibrosis of the vessel wall initiates an inflammatory response with the activation of microglia and recruitment of circulating macrophages, which release toxic molecules that breakdown the extracellular matrix and disrupt the blood–brain barrier (BBB). While this results in damage to the tissues surrounding the blood vessel; some of these changes may be involved in an attempt to repair the vessel.

Small vessel disease as a cause of dementia can progress slowly with a wide variety of symptoms. In addition, in the early stages it overlaps with neurodegenerative diseases, such as AD, further confounding early diagnosis. It is now recognized that small vessel disease is the main cause of VCID being more prevalent than large vessel disease. Small vessel disease has two forms: (1) a small stroke form that has one or more lacunar infarcts, principally in the basal ganglia were the caliber of the vessels abruptly change, leading to excess pressure on the smaller vessels in the basal ganglia, predisposing them to damage to the vessel walls; and (2) hypoxic hypoperfusion affecting mainly the distal parts of the vascular tree, the watershed regions. While lacunar strokes are generally multiple and occur randomly, the hypoxia due to the damaged vessels primarily affects the white matter, leading to the commonly observed WMHs on MRI. Small vessel disease more characteristically results in changes in the white matter rather than the gray matter, which is the most common site for large strokes.

Blood vessels and white matter lesions

Patterns of WMHs on FLAIR are related to the distribution of blood vessels. Vessels penetrate into the brain from either the surface or the base of the brain (Fig. 4.2) (Duvernoy et al., 1981). Passing through the cortex, the vessels branch at the junction of the gray/white matter: the transition from the large caliber vessel to the branched smaller vessels creates a strain similar to that occurring in the basal ganglia as the large vessels from the neck branch into the smaller vessels supplying the basal ganglia. Some the vessels at the gray/white junction form corkscrews: this makes the junction of the cortical gray matter and the underlying white matter a region vulnerable to both reduced flow and increased BBB permeability; this can be seen in the increase in FLAIR signal at the gray/white junction (Fig. 4.2C).

The end of an arterial circulation is vulnerable to a reduction in blood flow. Such a region is called a "watershed", indicating that it is the region most vulnerable when blood flow is lost. The central regions of the white matter that form the corpus callosum and internal and external capsule are watershed areas for CBF; when the flow is reduced they are particularly vulnerable to hypoxic hypoperfusion. Blood flow to the deep white matter comes from the neck through the

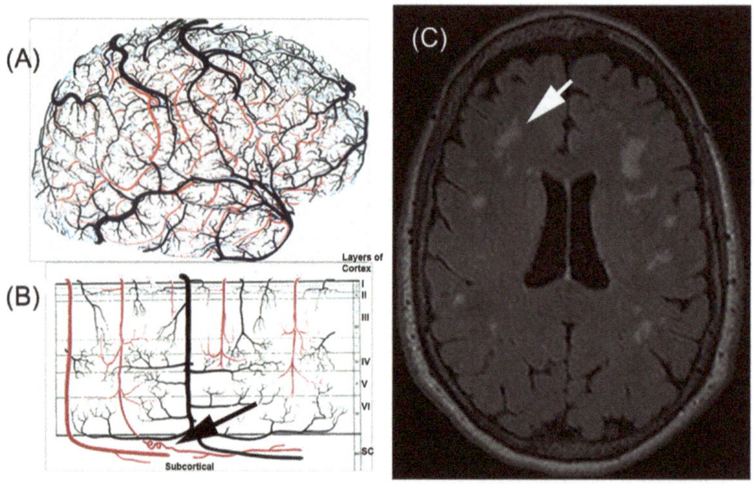

FIGURE 4.2 Increased in WMH FLAIR signal at the gray/white subcortical junction. Blood vessels in human brain. (A) Brain vasculature on the surface of the brain. Red plastic fills the arteries and blue plastic the veins. There are no anastomoses of the arteries and the veins since the arteries are end arteries. (B) This schematic drawing illustrates that the arteries pass from the surface to the deep white matter, traversing the VI layers of the cortex. Arterioles give off branches as the arteries pass through the cortex (*black arrow*). The capillaries that join the arterioles and the veins are not seen. (C) A FLAIR image from a patient with Binswanger's disease that shows the subcortical pattern of injury (*white arrow*). *FLAIR*, Fluid attenuated inversion discovery; *WMH*, white matter hyperintensities. *From Duvernoy, H. M., Delon, S., & Vannson, J. L. (1981). Cortical blood vessels of the human brain.* Brain Research Bulletin, 7*(5), 519–579.*

arteries at the base of the brain and from the cortical surface; these regions are perfused by end arteries and lack sufficient collateral circulation to maintain flow. This vulnerability explains the patterns of white matter injury seen on MRI.

Midlife hypertension and white matter injury

Increased systolic blood pressure is associated with clinically silent brain injuries such as increased brain atrophy, gray-matter atrophy, white-matter injury including WMHs, and clinically asymptomatic infarcts. In a study of changes in MRI scans in a third generation of a large cohort of subjects in the Framingham Study with information available on midlife blood pressure. Age was associated with decreased fractional anisotropy (FA) and increased mean diffusivity (MD) in cerebral white-matter voxels; systolic blood pressure was associated with injury to white-matter microstructure and regional gray matter later in life in young adults. Injury worsened continuously as systolic blood pressure increased and was significantly different between normotensive and prehypertensive individuals in some regions; subtle vascular brain injury develops insidiously during life, with discernible effects even in young adults, emphasizing the need for early and optimum control of blood pressure (A., 1997; Maillard et al., 2012). The presence of hypertension in midlife increases brain aging by 7 years (Fig. 4.3).

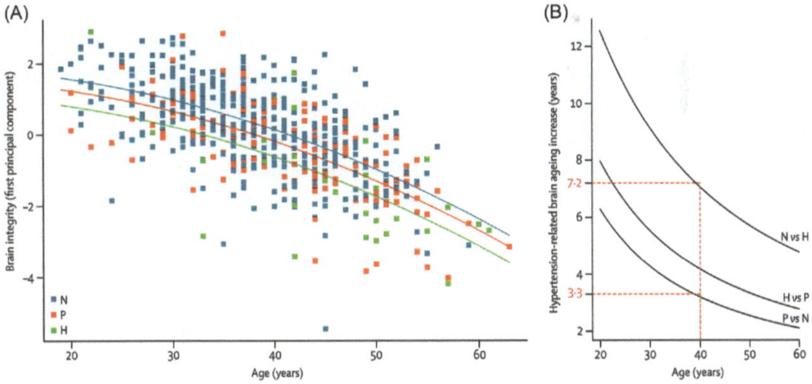

FIGURE 4.3 Impact of midlife hypertension on white matter microstructure. Regression curves relating brain integrity as expressed by the first principal component as a function of the hypertension category and age of the individual (A) and the difference in brain ageing increase between hypertension categories according to age (B) Red dotted lines indicate brain integrity of a 40-year-old by hypertensive status; brain integrity of a 40-year-old prehypertensive individual corresponded to that of a normotensive individual 3.3 years older, but this difference was 7.2 years for hypertensive individuals. *H*, Hypertensive; *N*, normotensive; *P*, prehypertensive. *From Maillard, P., Seshadri, S., Beiser, A., Himali, J. J., Au, R., Fletcher, E., Carmichael, O., Wolf, P. A., & DeCarli, C. (2012). Effects of systolic blood pressure on white-matter integrity in young adults in the Framingham Heart Study: A cross-sectional study.* The Lancet Neurology, *11(12), 1039–1047. https://doi.org/10.1016/S1474-4422(12)70241-7.*

Conversion of normal white matter to WMHs was correlated with lower baseline FA and higher baseline FLAIR intensity: WMH result from a continuous process of WM degeneration with time. Stepwise decreases in WM integrity as measured by both diffusion tensor imaging (DTI) and FLAIR were independently associated with stepwise increases in WMH risk, emphasizing that these modalities may provide complementary information for understanding the time course of aging-associated WM degeneration (Maillard et al., 2013).

Blood vessels and sites of selective vulnerability

Blood supply to the deep white matter comes from the leptomeningeal border zone; medullary arterioles wind their way up 4 cm before arriving at their ultimate destination (Moody et al., 1995). In addition, there are alterations in the venous system that occur with age: periventricular venous collagenosis is a mural disease of the periventricular veins. These changes in the arterioles and periventricular veins increase with aging and lead to tortuosity and multiplications of the vessels. Since these are end arteries, these alterations cause hypoperfusion of the watershed regions of the deep white matter as shown (Fig. 4.4) (Moody et al., 1990).

Brain arteries are end arteries with few anastomoses

Injury to the brain is highly dependent on the cerebral blood supply. Embryology of the brain vessels has shown that the vessels penetrate into deeper structures from the surface and the base of the brain; this pattern makes the deeper structures at the end of the system more vulnerable to reduced CBF. In some regions the vessel caliber makes a large change, going from a relatively large artery to a much smaller one; this can create pressure on the smaller vessel.

The "vascular vulnerability" of the hippocampus is due to the relatively small number of capillary anastomoses; this gives it a limited ability to withstand hypoperfusion. The long tangential course of arteries and veins in the hippocampus differs from that of the isocortical vessels, which are usually short and perpendicular to the surface with a palisade aspect. The particular vulnerability of hippocampal tissue to anoxia may be explained by penetration of the large intrahippocampal arteries at a perpendicular angle and by their long course with few anastomoses (Fig. 4.5) (Spallazzi et al., 2019).

BBB permeability is increased in white matter in patients with Binswanger's disease. Patterns of BBB permeability can be identified by dynamic contrast-enhanced MRI (DCEMRI): increased BBB permeability in white matter are in the same regions that show increased FLAIR signal; three regions that show increased BBB permeability are (1) subcortical,

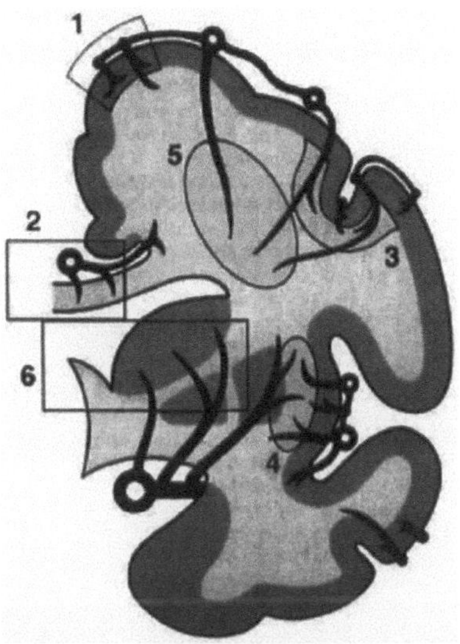

FIGURE 4.4 **Blood vessel supply to the deep white matter.** Scheme of arteriolar and arterial supply to cerebrum as viewed in coronal plane. Over the surface of the brain the leptomeningeal arteries are connected, but once these vessels penetrate the brain, they are end arteries. When normal, these intraparenchymal vessels do not have arteriolar-arteriolar or arterio-venous shunts. 1 = cortex; 2 = corpus callosum; 3 = subcortical U fibers; 4 = external capsule/claustrum/extreme capsule; 5 = centrum semiovale; 6 = basal ganglia and thalamus. Zones 3 and 4 have interdigitating arterioles whose parent vessels arise from different pial arteries, in some instances widely separated. (In zone 3, to cite the most extreme example, arterioles arising from anterior and middle cerebral artery systems might interdigitate with each other and share a capillary bed). *From Moody, D. M., Bell, M. A., & Challa, V. R. (1990). Features of the cerebral vascular pattern that predict vulnerability to perfusion or oxygenation deficiency: An anatomic study.* AJNR. American Journal of Neuroradiology, 11*(3), 431–439.*

(2) deep white matter, and (3) periventricular. WMHs develop in these hypoperfused regions; they have reduced CBF, particularly around the edges of the WMHs (Huisa et al., 2015). Increased permeability is found in regions outside the WMHs in "normal appearing white matter" which is an area without FLAIR signal.

Subtle changes in the permeability of the BBB occur as part of normal aging, subcortical ischemic vascular disease (SIVD), and AD; these can be detected with DCEMRI (Raja et al., 2018). Patients with chronic small vessel disease had increased BBB in normal appearing white matter, WMHs, and gray matter (Zhang et al., 2017). Low baseline CBF and FA, and high MD, were independently associated with new periventricular WMH voxels, with MD being the best predictor of WMH growth. CBF and DTI measures independently predict WMH growth over time. DTI is a more sensitive predictor of WMH growth than CBF, with WMH progression likely due to myelin injury secondary to low perfusion. Findings support the use of MD as a

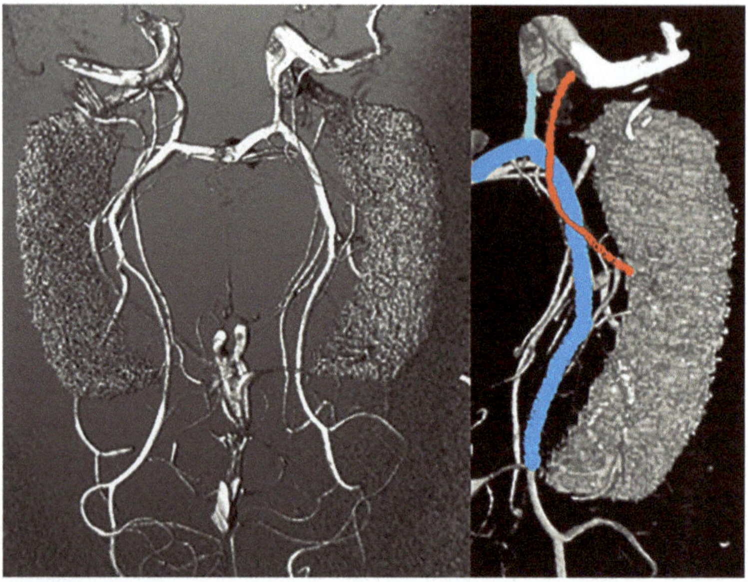

FIGURE 4.5 **On the left: a TOF MRA reconstruction with the hippocampal mask; on the right: main vessels involved in the hippocampal supply; posterior cerebral artery (light blue), anterior choroidal artery (red), posterior communicating artery (cyan).** *From Spallazzi, M., Dobisch, L., Becke, A., Berron, D., Stucht, D., Oeltze-Jafra, S., Caffarra, P., Speck, O., & Düzel, E. (2019). Hippocampal vascularization patterns: A high-resolution 7 Tesla time-of-flight magnetic resonance angiography study.* NeuroImage: Clinical, 21, *101609. https://doi.org/10.1016/j.nicl.2018.11.019.*

sensitive marker of normal appearing WM vulnerability in future trials aimed at preserving WM integrity (Promjunyakul et al., 2018).

DTI is another MRI method to show ischemic damage in the white matter that is more readily implemented on most MRI scanners than ^{1}H-MRS. Injury to the white matter alters the movement of water. Abnormalities in the white matter signal are an important biomarker to indicate vascular disease that can be used to identify patients with SIVD and combined with other biomarkers, especially for AD, it can be used for identification of mixed dementia.

In pathological surveys of patients with dementia, the small vessel disease is the most common variety (Gold et al., 2002; Schneider et al., 2007; Toledo et al., 2013). White matter is the major site of pathological changes associated with small vessel disease. Four patterns of white matter injury are revealed on FLAIR MRI: Subcortically in the region below the gray matter at the gray-white interface, in the centrum semiovale between the cortex and the ventricles, and surrounding the cerebral ventricles. These regions are poorly perfused due to the structure of the brain blood supply: they are regions between major vascular supplies, making them vulnerable to reductions in CBF (De Reuck, 1971; Moody et al., 1995).

Monogenic forms of white matter disease

CADASIL and CARASIL: autosomal dominant and recessive small vessel disease

Several genetic diseases related to a single-gene defect result in white matter disease: these include cerebral autosomal dominant arteriopathy with subcortical infarcts and leukoencephalopathy (CADASIL) and cerebral autosomal recessive arteriopathy with subcortical infarcts and leukoencephalopathy (CARASIL). These are rare forms of white matter disease with CADASIL being the most common on these monogenic diseases. CADASIL affects the small blood vessels in the white matter of the brain. Patients symptoms include migraine headaches, and they have multiple small strokes. Eventually, they develop dementia. There are a wide variety of symptoms with differing age of onset. Other symptoms include white matter lesions throughout the brain, cognitive deterioration, seizures, vision problems, and psychiatric problems such as severe depression and changes in behavior and personality. Individuals may also be at higher risk of heart attacks. Symptoms and disease onset vary widely, with signs typically appearing in the mid-30s. Some individuals may not show signs of the disease until later in life. CADASIL is caused by a mutation in the gene for NOTCH3. Inheritance is autosomal dominant. Treatment is supportive and limited to controlling symptoms. Most people with CADASIL become bed-ridden and develop dementia over time. Life expectancy is also reduced from lung and heart diseases in people with CADASIL. Controlling vascular risk factors is an important part of CADASIL management. The acronym CADASIL was suggested in 1993 to designate and characterize a hereditary disease of small cerebral arteries that affects middle-aged adults and leads to disability and dementia (Chabriat et al., 2009).

Cerebral autosomal recessive arteriopathy with subcortical infarcts and leukoencephalopathy (CARASIL) is a familial autosomal recessive vascular disorder characterized by nonhypertensive cerebral small vessel disease with early adulthood onset, subcortical infarcts, progressive motor and cognitive impairment, alopecia, and spondylosis. CARASIL is a single-gene disorder caused by mutations in the HtrA serine peptidase/protease 1 (HTRA1) gene. This gene encodes HTRA1, repressing the inhibition of TGF-β family signaling (Hara et al., 2009). Intense arteriolosclerosis without granular osmiophilic materials, associated with fibrous intimal proliferation, hyaline degeneration of the media, thickening and splitting of the internal elastic lamina, and concentric narrowing of the lumen, are the histopathologic characteristic features. CARASIL is extremely rare, with most cases reported in Asia, mainly Japan (Yanagawa et al., 2002).

References

A., S. D. (1997). Brain infarction and the clinical expression of Alzheimer disease. The Nun Study. *JAMA: The Journal of the American Medical Association*, 813–817. Available from https://doi.org/10.1001/jama.277.10.813.

Benson, D. F., Marsden, C. D., & Meadows, J. C. (1974). The amnesic syndrome of posterior cerebral artery occlusion. *Acta Neurologica Scandinavica, 50*(2), 133–145.

Blessed, G., Tomlinson, B. E., & Roth, M. (1968). The association between quantitative measures of dementia and of senile change in the cerebral grey matter of elderly subjects. *The British Journal of Psychiatry : The Journal of Mental Science, 114*(512), 797–811. Available from https://doi.org/10.1192/bjp.114.512.797.

Buratti, L., Balucani, C., Viticchi, G., Falsetti, L., Altamura, C., Avitabile, E., Provinciali, L., Vernieri, F., & Silvestrini, M. (2014). Cognitive deterioration in bilateral asymptomatic severe carotid stenosis. *Stroke: A Journal of Cerebral Circulation, 45*(7), 2072–2077. Available from https://doi.org/10.1161/STROKEAHA.114.005645.

Caplan, L. R., & Schoene, W. C. (1978). Clinical features of subcortical arteriosclerotic encephalopathy (Binswanger disease). *Neurology, 28*(12), 1206–1215. Available from https://doi.org/10.1212/wnl.28.12.1206.

Chabriat, H., Joutel, A., Dichgans, M., Tournier-Lasserve, E., & Bousser, M. G. (2009). CADASIL. *The Lancet Neurology, 8*(7), 643–653. Available from https://doi.org/10.1016/S1474-4422(09)70127-9.

Chui, H. C., Victoroff, J. I., Margolin, M., Jagust, W., Shankle, R., & Katzman, R. (1992). Criteria for the diagnosis of ischemic vascular dementia proposed by the state of California Alzheimer's disease diagnostic and treatment centers. *Neurology, 42*(3), 473–480. Available from https://doi.org/10.1212/wnl.42.3.473.

De Reuck, J. (1971). The human periventricular arterial blood supply and the anatomy of cerebral infarctions. *European Neurology, 5*(6), 321–334. Available from https://doi.org/10.1159/000114088.

Duvernoy, H. M., Delon, S., & Vannson, J. L. (1981). Cortical blood vessels of the human brain. *Brain Research Bulletin, 7*(5), 519–579.

Gold, G., Bouras, C., Canuto, A., Bergallo, M. F., Herrmann, F. R., Hof, P. R., Mayor, P. A., Michel, J. P., & Giannakopoulos, P. (2002). Clinicopathological validation study of four sets of clinical criteria for vascular dementia. *American Journal of Psychiatry, 159*(1), 82–87. Available from https://doi.org/10.1176/appi.ajp.159.1.82.

Hachinski, V., Iadecola, C., Petersen, R. C., Breteler, M. M., Nyenhuis, D. L., Black, S. E., Powers, W. J., DeCarli, C., Merino, J. G., Kalaria, R. N., Vinters, H. V., Holtzman, D. M., Rosenberg, G. A., Wallin, A., Dichgans, M., Marler, J. R., & Leblanc, G. G. (2006). National Institute of Neurological Disorders and Stroke-Canadian Stroke Network vascular cognitive impairment harmonization standards. *Stroke: A Journal of Cerebral Circulation, 37*(9), 2220–2241.

Hachinski, V. C., Iliff, L. D., Zilhka, E., Du Boulay, G. H., McAllister, V. L., Marshall, J., Russell, R. W., & Symon, L. (1975). Cerebral blood flow in dementia. *Archives of Neurology, 32*(9), 632–637.

Hachinski, V. C., Lassen., & Marshall, J. (1974). Multi-infarct dementia. A cause of mental deterioration in the elderly. *Lancet, 2*, 207–210.

Hachinski, V. C., Potter, P., & Merskey, H. (1987). Leuko-Araiosis. *Archives of Neurology, 44*(1), 21–23. Available from https://doi.org/10.1001/archneur.1987.00520130013009.

Hara, K., Shiga, A., Fukutake, T., Nozaki, H., Miyashita, A., Yokoseki, A., Kawata, H., Koyama, A., Arima, K., Takahashi, T., Ikeda, M., Shiota, H., Tamura, M., Shimoe, Y., Hirayama, M., Arisato, T., Yanagawa, S., Tanaka, A., Nakano, I., ... Onodera, O. (2009). Association of HTRA1 mutations and familial ischemic cerebral small-vessel disease. *The New England Journal of Medicine*, *360*(17), 1729–1739. Available from https://doi.org/10.1056/NEJMoa0801560.

Huisa, B. N., Caprihan, A., Thompson, J., Prestopnik, J., Qualls, C. R., & Rosenberg, G. A. (2015). Long-term blood-brain barrier permeability changes in Binswanger disease. *Stroke: A Journal of Cerebral Circulation*, *46*(9), 2413–2418. Available from https://doi.org/10.1161/STROKEAHA.115.009589.

Huisa, B. N., & Rosenberg, G. A. (2014). Binswanger's disease: Toward a diagnosis agreement and therapeutic approach. *Expert Review of Neurotherapeutics*, *14*(10), 1203–1213. Available from https://doi.org/10.1586/14737175.2014.956726.

Hunt, A. L., Orrison, W. W., Yeo, R. A., Haaland, K. Y., Rhyne, R. L., Garry, P. J., & Rosenberg, G. A. (1989). Clinical significance of MRI white matter lesions in the elderly. *Neurology*, *39*(11), 1470–1474.

Iadecola, C., Duering, M., Hachinski, V., Joutel, A., Pendlebury, S. T., Schneider, J. A., & Dichgans, M. (2019). Vascular cognitive impairment and dementia: JACC scientific expert panel. *Journal of the American College of Cardiology*, *73*(25), 3326–3344. Available from https://doi.org/10.1016/j.jacc.2019.04.034.

Kaczmarz, S., Göttler, J., Petr, J., Hansen, M. B., Mouridsen, K., Zimmer, C., Hyder, F., & Preibisch, C. (2021). Hemodynamic impairments within individual watershed areas in asymptomatic carotid artery stenosis by multimodal MRI. *Journal of Cerebral Blood Flow and Metabolism : Official Journal of the International Society of Cerebral Blood Flow and Metabolism*, *41*(2), 380–396. Available from https://doi.org/10.1177/0271678X20912364.

Karanth, S., Nelson, P. T., Katsumata, Y., Kryscio, R. J., Schmitt, F. A., Fardo, D. W., Cykowski, M. D., Jicha, G. A., Van Eldik, L. J., & Abner, E. L. (2020). Prevalence and clinical phenotype of quadruple misfolded proteins in older adults. *JAMA Neurology*, *77*(10), 1299–1307. Available from https://doi.org/10.1001/jamaneurol.2020.1741.

Kim, D., Yang, P.-S., Lip, G. Y. H., & Joung, B. (2020). Atrial fibrillation increases the risk of early-onset dementia in the general population: Data from a population-based cohort. *Journal of Clinical Medicine*, *9*(11). Available from https://doi.org/10.3390/jcm9113665.

Kim, D., Yang, P.-S., Sung, J.-H., Jang, E., Yu, H. T., Kim, T.-H., Uhm, J.-S., Kim, J.-Y., Pak, H.-N., Lee, M.-H., Lip, G. Y. H., & Joung, B. (2020). Less dementia after catheter ablation for atrial fibrillation: A nationwide cohort study. *European Heart Journal*, *41*(47), 4483–4493. Available from https://doi.org/10.1093/eurheartj/ehaa726.

Kinkel, W. R., Jacobs, L., Polachini, I., Bates, V., & Heffner, R. R. (1985). Subcortical arteriosclerotic encephalopathy (Binswanger's disease). Computed tomographic, nuclear magnetic resonance, and clinical correlations. *Archives of Neurology*, *42*(10), 951–959.

Lee, Z. X., Ang, E., Lim, X. T., & Arain, S. J. (2021). Association of risk of dementia with direct oral anticoagulants vs warfarin use in patients with non-valvular atrial fibrillation: A systematic review and meta-analysis. *Journal of Cardiovascular Pharmacology*, *77*(1), 22–31. Available from https://doi.org/10.1097/FJC.0000000000000925.

Maillard, P., Carmichael, O., Harvey, D., Fletcher, E., Reed, B., Mungas, D., & DeCarli, C. (2013). FLAIR and diffusion MRI signals are independent predictors of white matter hyperintensities. *AJNR. American Journal of Neuroradiology*, *34*(1), 54–61. Available from https://doi.org/10.3174/ajnr.A3146.

Maillard, P., Seshadri, S., Beiser, A., Himali, J. J., Au, R., Fletcher, E., Carmichael, O., Wolf, P. A., & DeCarli, C. (2012). Effects of systolic blood pressure on white-matter integrity in young adults in the Framingham Heart Study: A cross-sectional study. *The Lancet Neurology*, *11*(12), 1039–1047. Available from https://doi.org/10.1016/S1474-4422(12)70241-7.

Moody, D. M., Bell, M. A., & Challa, V. R. (1990). Features of the cerebral vascular pattern that predict vulnerability to perfusion or oxygenation deficiency: An anatomic study. *AJNR. American Journal of Neuroradiology*, *11*(3), 431–439.

Moody, D. M., Brown, W. R., Challa, V. R., & Anderson, R. L. (1995). Periventricular venous collagenosis: Association with leukoaraiosis. *Radiology*, *194*(2), 469–476. Available from https://doi.org/10.1148/radiology.194.2.7824728.

Norling, A. M., Marshall, R. S., Pavol, M. A., Howard, G., Howard, V., Liebeskind, D., Huston, J., Lal, B. K., Brott, T. G., & Lazar, R. M. (2019). Is hemispheric hypoperfusion a treatable cause of cognitive impairment? *Current Cardiology Reports*, *21*(1), 4. Available from https://doi.org/10.1007/s11886-019-1089-9.

Pantoni, L. (2010). Cerebral small vessel disease: From pathogenesis and clinical characteristics to therapeutic challenges. *The Lancet Neurology*, *9*(7), 689–701. Available from https://doi.org/10.1016/S1474-4422(10)70104-6.

Promjunyakul, N.-O., Dodge, H. H., Lahna, D., Boespflug, E. L., Kaye, J. A., Rooney, W. D., & Silbert, L. C. (2018). Baseline NAWM structural integrity and CBF predict periventricular WMH expansion over time. *Neurology*, *90*(24), e2119–e2126. Available from https://doi.org/10.1212/WNL.0000000000005684.

Raja, R., Rosenberg, G. A., & Caprihan, A. (2018). MRI measurements of blood-brain barrier function in dementia: A review of recent studies. *Neuropharmacology*, *134*(Pt B), 259–271. Available from https://doi.org/10.1016/j.neuropharm.2017.10.034.

Rosenberg, G. A., Kornfeld, M., Stovring, J., & Bicknell, J. M. (1979). Subcortical arteriosclerotic encephalopathy (Binswanger): Computerized tomography. *Neurology*, *29*(8), 1102–1106.

Schneider, J. A., Arvanitakis, Z., Bang, W., & Bennett, D. A. (2007). Mixed brain pathologies account for most dementia cases in community-dwelling older persons. *Neurology*, *69*(24), 2197–2204. Available from https://doi.org/10.1212/01.wnl.0000271090.28148.24.

Snyder, H. M., Corriveau, R. A., Craft, S., Faber, J. E., Greenberg, S. M., Knopman, D., Lamb, B. T., Montine, T. J., Nedergaard, M., Schaffer, C. B., Schneider, J. A., Wellington, C., Wilcock, D. M., Zipfel, G. J., Zlokovic, B., Bain, L. J., Bosetti, F., Galis, Z. S., Koroshetz, W., & Carrillo, M. C. (2015). Vascular contributions to cognitive impairment and dementia including Alzheimer's disease. *Alzheimer's & Dementia : The Journal of the Alzheimer's Association*, *11*(6), 710–717. Available from https://doi.org/10.1016/j.jalz.2014.10.008.

Spallazzi, M., Dobisch, L., Becke, A., Berron, D., Stucht, D., Oeltze-Jafra, S., Caffarra, P., Speck, O., & Düzel, E. (2019). Hippocampal vascularization patterns: A high-resolution 7 Tesla time-of-flight magnetic resonance angiography study. *NeuroImage: Clinical*, *21*, 101609. Available from https://doi.org/10.1016/j.nicl.2018.11.019.

Toledo, J. B., Arnold, S. E., Raible, K., Brettschneider, J., Xie, S. X., Grossman, M., Monsell, S. E., Kukull, W. A., & Trojanowski, J. Q. (2013). Contribution of cerebrovascular disease in autopsy confirmed neurodegenerative disease cases in the National Alzheimer's Coordinating Centre. *Brain*, *136*(9), 2697–2706. Available from https://doi.org/10.1093/brain/awt188.

Yanagawa, S., Ito, N., Arima, K., & Ikeda, S. (2002). Cerebral autosomal recessive arteriopathy with subcortical infarcts and leukoencephalopathy. *Neurology*, *58*(5), 817–820.

Zhang, C. E., Wong, S. M., van de Haar, H. J., Staals, J., Jansen, J. F. A., Jeukens, C. R. L. P. N., Hofman, P. A. M., van Oostenbrugge, R. J., & Backes, W. H. (2017). Blood-brain barrier leakage is more widespread in patients with cerebral small vessel disease. *Neurology*, *88*(5), 426–432. Available from https://doi.org/10.1212/WNL.0000000000003556.

Chapter 5

Neuroinflammation

Classic definition of inflammation

The effects of inflammation have been known since ancient times because they are easily visible. They include redness (*rubor*), swelling (*tumor*), heat (*calor*), pain (*dolor*) and loss of function (*functio laesa*). After an early description by Celsus (30–38 BCE), the list was completed by Galen (130–200 AD). The inflammatory changes, which were seen mainly in the extremities and the skin, were related to infections. Up until recently, the emphasis was placed on tissue injury. However, beginning in the in the 19th century, the importance of inflammation in the body's defensive and healing processes was recognized (Punchard et al., 2004). While the effects of inflammation were obvious in many parts of the body, the impact of inflammation in the brain were mainly reserved for serious infections, such as bacterial meningitis and the concept of chronic inflammation was rarely considered in the nervous system. The classic effects of inflammation seen in other organs such as increased movement of blood through dilated vessels that resulted in redness, and increased vascular permeability, leading to extravasation of fluid and inflammatory cells into the tissue, causing swelling, in the brain these could only be observed at autopsy. While prolonged inflammatory responses that caused pain and loss of function were commonly diagnosed in the connective tissue, particularly in the joints, awareness of the complex changes that were taking place in the brain were not fully appreciated until the modern era of neuroimaging. Today, the term inflammation is used in a very broad sense ranging from infection by invading organisms to more subtle changes that result from a response of the central nervous system (CNS) to tissue injury from trauma, ischemia, and autoimmunity. Chronically, the effects of inflammation can be more subtle with activation of the inflammatory cells of the brain and the recruitment of inflammatory cells from the systemic circulation. While the adverse effects of chronic inflammation are well recognized in the body, such as inflammation of joints in arthritis and inflammation of the lungs in asthma, changes in the brain are more subtle, particularly damage to the blood vessels as a results of hypertension and diabetes, and can go unrecognized for many years until reversal of the damage is no longer possible.

Neuroinflammation in Vascular Dementia. DOI: https://doi.org/10.1016/B978-0-12-823455-6.00007-9

With the major advances in brain imaging and a deeper understanding of the molecular processes taking place in the brain, the concept of inflammation has been expanded beyond the original meaning to be possibly present in all aspects of brain injury and repair. This broader use has diluted the relevance, and some constraint is needed to restore the word inflammation to a more limited and useful meaning. This chapter will focus on the consequences of acute and chronic inflammation and on understanding the stages in this process, which is critical in identifying the underlying disease processes and in planning treatments. Some of the events are present early and contribute to tissue damage, while others begin later, promoting healing. As more is learned, it is possible to understand that early stages of inflammation worsen injury, while later stages participate in repair. This dual function of inflammation has made the search for treatments more difficult; it is critical to know the stages of inflammation to be able to control the acute injury phase while not interfering later with repair. Thus, it is important not only to think of the molecules that the treatment is blocking, but also to determine when they begin to promote repair.

Several major classes of molecules are involved in each of the stages: initially there is release of highly toxic substances that contribute to tissue destruction and removal of damaged tissue; at this stage inflammatory cells release proteases, free radicals, and cytokines, which are critical at multiple stages. Proteases begin the process of clearing away the debris by breaking down the extracellular matrix, attacking the cell walls, and activating cells in the immune response. Apoptosis leads to cell death and autophagy removes and recycles cellular debris to conserve essential cellular components with free radicals of nitrogen and oxygen contributing to this process; toxic free radicals are released by invading macrophages and endogenous microglia. Neutrophils, T-cells, B-cells, and circulating macrophages are recruited from the systemic circulation.

The majority of studies have focused on acute inflammation, delineating the molecular cascades involved in the early stages of injury. Understanding the consequences of chronic inflammation on the brain has evolved more slowly. This is particularly true of diseases of aging, where the immune cells create an adverse microenvironment. Aging releases the immune system control mechanisms in a process that is called inflammaging. Neuroinflammation is now recognized as an important component of diseases of aging. The role of genetics and the molecular cascades involved in other chronic inflammatory conditions are recognized as important in brain disorders: opening new areas of research into the molecular events involved in these disorders.

Inflammation in chronic hypoxic/ischemic disease

Acute lack of oxygen leads to cell death, while chronic hypoxia, which can be tolerated over extended periods, initiates molecular cascades that more slowly affect cellular function. When the brain is subjected to low-grade hypoxia, molecular events cause tissue damage. An example of chronic hypoxia is the mountain climber without supplemental oxygen. Another more common example is the intermittent hypoxia of sleep apnea. Chronic hypoxia induces molecular changes with the goal of increasing the delivery of oxygen to brain cells by producing red blood cells (erythropoiesis) and blood vessel growth (angiogenesis). The combination of slowly evolving tissue destruction with regeneration of tissue has been a major hurdle in devising effective treatments that block one without impeding the other. Dementia is caused by a heterogeneous group of diseases that progress slowly to cell death, although through different pathways. In Alzheimer's and related diseases, misfolded proteins accumulate in amyloid plaques around blood vessels and lead to an increase in phosphorylated tau in the neuron that is the cause of cell death. additional misfolded proteins accumulate with aging, adding to inflammation. As opposed to the direct toxicity of abnormal proteins, blood vessels are adversely affected by age and damaged by vascular risk factors. The blood vessel disease leads to restriction of cerebral blood flow, depriving the brain of oxygen and nutrients, which kills brain cells. The misfolded proteins and vascular disease represent dual pathways that begin as separate disease processes, but ultimately come together to produce mixed dementia, which is the most common form of dementia. The inevitable progression of two disease processes provides an opportunity to determine the stage of each and to tailor therapy to the appropriate stage. Biomarkers can identify the pathways involved, aid in classifying the underlying disease processes, and be used as surrogate endpoints in treatment trials.

Consequences of reduced cerebral blood flow

Brain tissue is dependent on a continuous supply of blood to deliver glucose and oxygen to the tissue; to meet this constant need, the control of cerebral blood flow occurs at multiple levels (Iadecola & Nedergaard, 2007). Vasoconstriction and dilation are under both chemical and neural control. Normal vessels are pliable and able to respond to a variety of conditions. The blood flow is autoregulated to remain constant over a wide blood pressure range, which assures that changing position from sitting to standing, results in vasodilation. One of the consequences of aging is the increase of rigidity of the blood vessels impeding their ability to respond to postural changes. Chronic hypertension causes a narrowing of the vessel lumen and fibrosis of the outer vessel wall. The result is both a reduction of cerebral blood flow and loss of responsiveness. In some diseases, the blood vessels have a similar effect without hypertension. An example of a nonhypertensive disease process that damages the blood vessels is the hereditary autosomal dominant disease, CADASIL, in which an abnormal protein is deposited in the vessel wall. Cerebral blood flow is sensitive to changes in the concentration of carbon dioxide; increased concentration of carbon dioxide cause vasodilation.

Chronic hypoxia is a serious consequence of reduced cerebral blood flow. When oxygen is low, there is the release of hypoxia inducible factor-1α (HIF-1α), which triggers expression of a number of molecules that participate in tissue injury and simultaneously begin the repair process (Kaelin et al., 2016). Hypoxia-inducible factors (HIFs) are transcription factors that respond to decreases in available oxygen in the cellular environment, or hypoxia. When brain cells sense a reduction in oxygen, there is a rapid molecular response that is controlled by HIF-1α. When oxygen is low, Von Hippel–Lindau tumor suppressor gene (VHL) binds to HIF-1α, activating genes directed at increasing the oxygen content of the brain. Under normal conditions in the presence of oxygen, HIF-1α is regulated by hydroxylation. When iron, 2-oxoglutarate, and oxygen are present, HIF is inactivated by HIF hydroxylases. Hydroxylation of HIF creates a binding site for pVHL (the protein product of the VHL gene). pVHL directs the polyubiquitylation of HIF-1α, ensuring that this protein will be degraded by the proteasome. In hypoxic conditions, HIF-1α subunits accumulate and bind to HIF-1β. This heterodimer of HIF is a transcription factor that activates genes that encode for proteins such as vascular endothelial growth factor (VEGF) the major protein involved in angiogenesis. Under hypoxic conditions, oxygen-requiring HIF hydroxylase enzymes, PHD1, 2, and 3 are inactivated and VHL does not bind HIF, allowing HIF to remain and activate the EPO gene (Kaelin et al., 2016). Hypoxia responsive element (HRE) is involved in the activation of a number of genes that are produced during hypoxia: some of these genes increase metabolism, others lead to growth of blood vessels, and erythropoietin increases red blood cells (Fig. 5.1) (Kaelin & Ratcliffe, 2008).

The central role of HIF-1α is key to both the injury and the repair processes because of the number of genes that it affects (Fig. 5.2). Hypertension, diabetes mellitus, and other diseases that damage blood vessels can initiate white matter injury by producing hypoxia. In addition, the injury to the vessel walls increases the risk of thrombosis and small strokes. If the blood flow to the watershed areas of the white matter is compromised by hypotension or by narrowing of the arteries, then hypoxia–ischemia induces inflammation. Hypoxia increases HIF-1α concentration, which binds with HIF-1β, and activates the hypoxia responsive element (HRE) in the target genes in the nucleus. This turns on cassettes of genes associated with injury such as the gene, *fur*, that produces the protein, furin, as well as VEGF and transforming growth factor-β (TGFβ); these angiogenic growth factors are important in repair. The convertase, furin, formed by the *fur* gene activates membrane type-metalloproteinase MT1-MMP, which activates MMP-2 as part of the trimolecular complex of MMP-2, MT1-MMP,

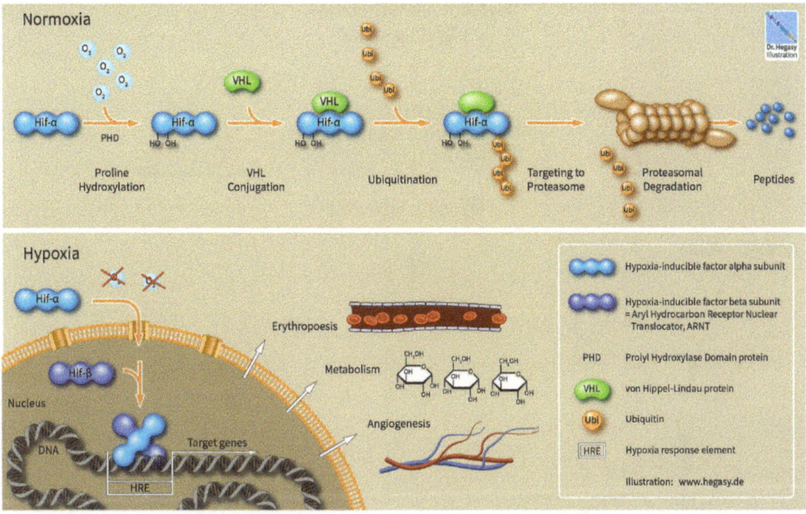

FIGURE 5.1 Hypoxia-inducible factor under normoxic and hypoxic conditions. Under well-oxygenated conditions HIF-1α becomes hydroxylated at one (or both) of two highly conserved prolyl residues located near the PHD family. Hydroxylation of either of these prolyl residues generates a binding site for the von Hippel-Lindau (pVHL) tumor suppressor protein, which is a component of a ubiquitin ligase complex. As a result, HIF-1α is polyubiquitylated and subjected to proteasomal degradation when oxygen is available. The PHD proteins belong to the Fe(II) and 2-oxoglutarate-dependent oxygenase superfamily, whose activity is absolutely dependent on oxygen. Accordingly, the rate of HIF hydroxylation is suppressed by hypoxia. Under low oxygen conditions, or in cells lacking functional pVHL, HIF-1α accumulates, dimerizes with an HIF-1β family member, translocates to the nucleus, and transcriptionally activates 100–200 genes, including genes involved in erythropoiesis, angiogenesis, autophagy, and energy metabolism. *HIF-1α*, Hypoxia-inducing factor 1α; *PHD*, prolyl hydroxylase domain. *From Kaelin, W. G., Ratcliffe, P. J., & Semenza, G. L. (2016). Pathways for oxygen regulation and homeostasis: The 2016 Albert Lasker basic medical research award.* JAMA, 316*(12), 1252–1253. https://doi.org/10.1001/jama.2016.12386).*

and tissue inhibitor to metalloproteinase-2 (TIMP-2). MMP-2 plays a major role in VCID by disrupting tight junction proteins and opening the blood brain barrier (BBB), leading to edema; in addition it causes bystander demyelination by attacking the myelinated axons. Thus, the combination of BBB disruption and myelin breakdown defines the major pathological changes seen in the brains of Binswanger patients with extensive white matter damage. Additionally, MMP-2 activates endothelin 1, which causes vasoconstriction through calcium metabolism in the small muscle. Vasoconstriction aggravates the hypoxic state. Conversely, on the repair side, VEGF and TGFβ activate angiopoietin 2, which acts through the secretion of MMPs to initiate angiogenesis and neurogenesis (Rosenberg, 2009).

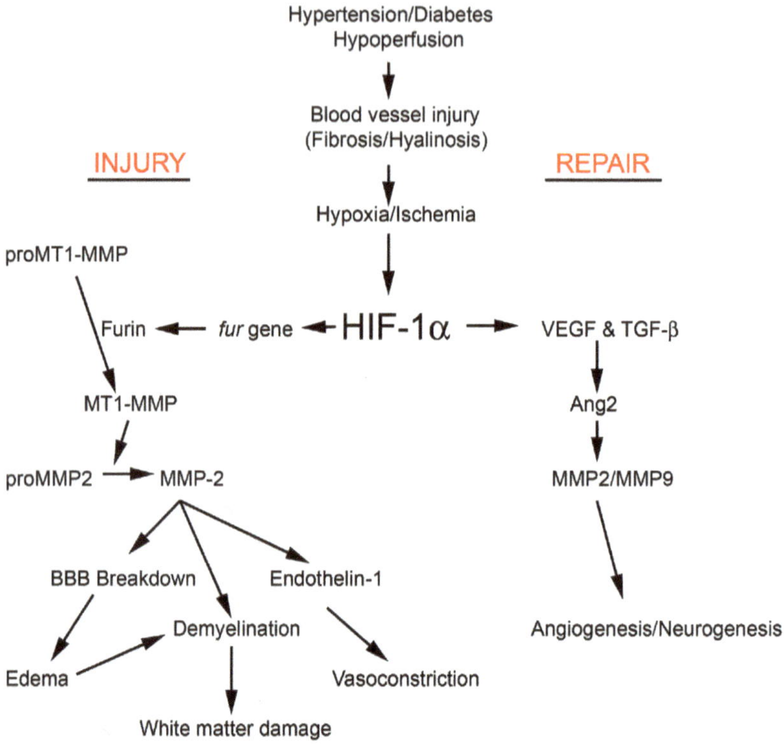

FIGURE 5.2 Schematic drawing of possible mechanisms for white matter injury in vascular cognitive impairment. Hypertension, diabetes mellitus, and other diseases that damage blood vessels can initiate white matter injury by reducing cerebral blood flow and increasing risk of thrombosis and small strokes. If the blood flow to the watershed areas of the white matter is compromised by hypotension or by narrowing of the arteries, then hypoxia–ischemia results, which induces inflammation. Hypoxia secondary to hypoperfusion increases HIF-1α concentration, which turns on cassettes of genes associated with injury such as *fur*, and increases expression of VEGF and TGF-β, which are important in repair. Furin leads to activation of MMP-2 through activation of MT1-MMP. MMP-2 can disrupt the tight junction proteins and open the BBB, leading to edema. Edema might also cause demyelination. Additionally, MMP-2 attacks myelin and activates endothelin 1, causing vasoconstriction through calcium metabolism in the small muscle. Vasoconstriction aggravates the hypoxic state. Conversely, on the repair side, VEGF and TGFβ activate angiopoietin 2, which acts through the secretion of MMPs to initiate angiogenesis and neurogenesis. *BBB*, Blood–brain barrier; *HIF-1α*, hypoxia-inducing factor 1α; *MMP*, matrix metalloproteinase; *MT*, membrane type; *TGFβ*, transforming growth factor β; *VEGF*, vascular endothelial growth factor.

Brain fluid interfaces

A series of interfaces between the brain and systemic circulation isolate brain tissue from the shock of changes in the concentration of electrolytes in blood. The first observation that the brain was protected from the systemic circulation was made by Paul Erhlich in 1885. He injected a water soluble dye into the peripheral circulation and showed that while organs in the body were stained with the dye, the brain remained unstained. The next insight came in 1913 when Edwin Goldmann injected dye into the CSF, showing that the brain was stained. These early observations formed the basis for the concept of a "blood–brain barrier" that allows the brain to function within a well-controlled environment separate from the systemic milieu of the periphery (Davson, 1976). Specialized cells separate at each of the sites of contact between the blood and cerebrospinal/interstitial fluids (CSF/ISF). At each of the interfaces are found tight junction proteins that seal the cells together: the endothelial cells with tight junction proteins, the choroid plexus apical cells, and the arachnoid cells on the surface of the brain and in the arachnoid granulations. Normally, these interfaces provide a barrier for most molecules except those that are lipid soluble or gases. Gap junctions allow for exchange of substances; they are found in the cells surrounding the ventricles. The presence of gap junctions allows substances injected intrathecally to move into brain tissue. Drugs that are excluded from the brain by the tight junctions at the interfaces can be delivered to the brain by intrathecal injection. Reese and Karnovsky, using electron microscopy with the electron dense molecule horseradish peroxidase (HRP) showed that the HRP was restricted from entering the brain by the presence of tight junctions between endothelial cells (Reese & Karnovsky, 1967). The large molecular weight, HRP, was blocked from crossing between the endothelial cells. Lack of pinocytotic vesicles is a characteristic feature of the endothelium of cerebral vessels; the low frequency of pinocytotic vesicles is an important differentiating feature of cerebral vessels as opposed to the large number of vesicles of the type that have been found to be associated with transport of materials across endothelia in other organs, such as the heart and muscle. Furthermore, no evidence was found of discharge of peroxidase on the contraluminal side of endothelium of cerebral vessels. Since there is evidence that vesicles in the vascular endothelia in cardiac and skeletal muscle transport materials by filling on the luminal side of the endothelial cells and discharging their contents into the perivascular spaces, the paucity of similar vesicles in brain and their failure to discharge their peroxidase on the contraluminal side of the vessels could be regarded as another manifestation of a blood–brain barrier. Cerebral vessels have micropinocytosis, serving some nutritional or other needs of the endothelial cell.

Glucose and amino acids are actively transported into the brain by carrier molecules in the endothelial cells. Removal of molecules from the brain is facilitated by active transport systems. Under normal conditions, cerebral vessels have few pinocytotic vesicles. However, when there is hypertension, the number of vesicles increases and they become an important mechanism for the movement of large molecules that are too big to move through the tight junctions, indicating that molecular size is important for the transport of substances from the blood to the brain. Interstitial fluid (ISF) is continuous with the CSF across the ependymal cells that line the ventricles. ISF-CSF function as the lymphatic system for the brain by removing the products of metabolism and delivering nutrients to the cells. Early work by Louis Weed and Harvey Cushing led to the concept that the ISF-CSF is the third circulation with blood being the first and lymph the second (Weed, 1914). Cerebral vessels secret ISF-CSF by the action of ATPase pumps on the surface of the endothelial cells. Fluid secretion creates oncotic pressure that pulls water into the brain extracellular space along an osmotic gradient. In the body, the lymphatics drain proteins that have left the blood back into the blood; this essential function prevents protein accumulation in the tissues where the increased oncotic pressure from the proteins in the tissues would cause edema. A different situation is found in the cerebral vessels, which restrict transport of most substances between the blood and the brain.

The separation between the systemic circulation and the brain cells is maintained by specialized proteins that self-assemble to form tight junctions in cerebral capillaries. Tight junction proteins prevent large protein molecules from entering the brain. Only lipid soluble substances and gases exchange readily between the blood and brain tissue. Energy is required for the function of the sodium-potassium ATPase exchange pumps; ISF is continuously produced at the capillaries with the fluid moving into the extracellular space along osmotic gradients created by the pumps on the abluminal surface. Similar mechanisms of fluid formation are found at the choroid plexus. Water movement is facilitated by the presence of a series of water channel forming molecules, the aquaporins; they form water transport channels in blood vessels and water passively moves along pressure and osmotic gradients (King et al., 2000). Fluid formed by brain capillaries and choroid plexus flows into the extracellular space as the ISF, eventually joining with the CSF in the ventricles and subarachnoid space to flow back into the blood. Arachnoid granulations drain the ISF-CSF back into the blood. ISF-CSF moves from the blood, through brain tissue in the extracellular space, entering the ventricles, and circulating back to the blood.

Cerebral extracellular fluids drain from brain to blood across the arachnoid villi and to lymph along certain cranial nerves (primarily olfactory) and spinal nerve root ganglia. The immune response to antigens (albumin or myelin basic protein) introduced into the CNS has been analyzed using a rat model with normal brain barrier permeability. The microinjection of antigen into brain or cerebrospinal fluid can be detected in the cervical lymph nodes where it elicits a humoral immune response with antibody production in cervical lymph nodes and spleen, and also affects cell-mediated immunity. These results support a highly regulated communication between the brain and the immune system in health that could be involved in immunological diseases (Cserr et al., 1992).

The CNS is the only organ system of the body lacking classically defined lymphatic vessels and tissues, and as a result has developed unique adaptations for waste clearance and fluid balance. Molecules of different molecular weights, such as albumin (69 kDa) and polyethylene glycols (900 Da and 4 kDa), are cleared from brain with nearly identical half-times of disappearance, leading to the conclusion that the efflux of ISF and its constituent solutes was by bulk fluid flow and not diffusion. There is a glial/lymphatic pathway for fluid transport in the brain that is responsible for movement of the CSF through the brain parenchyma. The explanation proposed for CSF movement into brain occurs within periarterial spaces and interchanges with ISF owing to the presence of aquaporin (AQP4) water channels in perivascular astrocyte end-foot processes; ISF then drains from brain within perivenous channels, again with AQP4 decreasing the resistance to fluid flow (Louveau et al., 2017). This movement along perivascular and perivenous pathways has a directional component; it is called bulk flow as opposed to the random multidirectional movement of diffusion. Bulk flow of ISF flows between white matter fiber tracts from the brain into the ventricles (Rosenberg et al., 1980). The pressure from continual formation of ISF at the capillary by the ATPase pumps is one possible mechanism for the propulsion of ISF. Another explanation is that CSF and ISF is driven by cerebral arterial pulsatility, which has been shown in rodent models, but the relevance to higher mammals is unclear. There is evidence that the removal of ISF is facilitated by sleep, suggesting that removal of toxic waste products, which occurs at night during sleep, may be a mechanism for removal of amyloid and other misfolded proteins with obvious implications for Alzheimer's disease (Reeves et al., 2020).

Extracellular matrix and the matrix metalloproteinases

The extracellular matrix is in the extracellular space and forms a thin layer of basal lamina around blood vessels. The basal lamina, which is thicker in other organs, such as the kidney, where it is called a basement membrane, consists of a mixture of collagens, laminin (glycoprotein), perlecan (heparan sulfate glycoprotein), and entactin (glycoprotein). These proteins can bind to each other to make a highly crosslinked extracellular matrix that functions as a molecular size and charge filter. Cerebral blood vessels differ from systemic vessels in several important features: they have few pinocytotic vesicles, carrier proteins that transport glucose and amino acids into the brain, and ATPase pumps in the abluminal surface that form the ISF. There are at least 40 different proteins composing the tight junctions. These proteins consist of both transmembrane and cytoplasmic proteins. The three major transmembrane proteins are occludin, claudins, and junction adhesion molecules. Tight junction proteins are located between endothelial cells, forming the major site of the blood–brain barrier (Hawkins & Davis, 2005).

Spread of metastatic cancer cells across the blood vessel to enter tissue is facilitated by matrix metalloproteinases (MMPs). The first MMP to be discovered was collagenase that was found in the regenerating tail of the tadpole (Brinckerhoff & Matrisian, 2002). Metastatic cancer cells use MMPs to breakdown the vessel wall, allowing for the cells to cross the blood vessels and invade tissue (Liotta & Stetler-Stevenson, 1990). In the brain the MMPs play a similar role, but they are produced in the brain during an inflammatory response and open the BBB from inside out. Once the MMPs are produced they attack the extracellular matrix and disrupt the tight junction proteins allowing fluid and entrained molecules to enter the brain.

The matrix metalloproteinases (MMPs) are a ubiquitous family of 26 matrix-degrading enzymes. They are classified into several families depending on their size and molecular action. Collagenases (MMP-1 and MMP-13) attack the various collagen molecules. Gelatinases (MMP-2 and MMP-9) are active in brain diseases with MMP-2 being a constitutive enzyme that is normally present in high concentrations in the brain, while MMP-9 is normally found at very low levels in the brain and is induced during inflammation and acute brain ischemia. MMP-2 is secreted by astrocytes that surround blood vessels in the foot processes. They may be involved in maintaining a normal basal lamina by preventing the basal lamina from overgrowing. MMP-9 is an inducible enzyme that is only present during acute inflammation. There are two stromelysins (MMP-3 and MMP-10). The membrane-type MMPs (MMP-14) activate MMP-2 (Rosenberg, 2009). All of the MMPs have a similar basic structure that includes a proteolytic zinc molecule. Transcription of the inflammatory MMPs is driven by the AP-1 and NF-kB sites in the genes (Fig. 5.3).

Role of the metalloproteinases in the disruption of the blood–brain barrier

Matrix metalloproteinases are both constitutive and inducible. Normally, brain has the constitutive MMP-2 and membrane-type metalloproteinase (MT-MMP): there is a low level of activity that allows MMP-2 to perform normal extracellular matrix maintenance. Because the MT-MMP, which is the activator of MMP-2, is close to the membrane, the action of MMP-2 is normally constrained to acting close to the membrane where its damage is contained. However, under hypoxic conditions the *fur* gene is formed, which leads to the production of the protein furin. Furin is a convertase that activates enzymes. In the presence of furin, MT-MMP is activated and, in turn, activates proMMP-2 to the active form. This is a constitutive process that occurs under normal circumstances; it most likely is important in maintaining a normal extracellular matrix. Hypoxia turns on a number of genes, including AP-1 and NF-κB: these two transcription factors are the main ones involved in the formation of the inflammatory MMPs, such as MMP-1, MMP-3 and MMP-9. In addition, MMP-3 activates the latent form of MMP-9. The membrane bound MMP is tethered to the cells around the blood

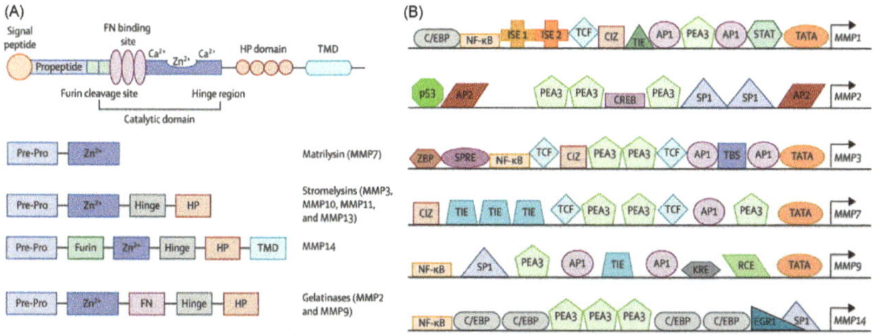

FIGURE 5.3 Family of matrix metalloproteinase molecules. MMP protein structure and binding sites on the promoters of the MMP genes (A) A signal peptide and the propeptide region form part of the cysteine switch, which folds over the zinc in the catalytic site and maintains a latent state. A cleavage site enables the proconvertase furin to activate MMP by cleaving the propeptide. An FN binding site is present in MMP2 and MMP9, connecting them with the basal lamina. The catalytic zinc site is present in all MMPs. A haemopexin domain is joined to the catalytic site by the hinge region. MMP14 (an MT-MMP) has a TMD. (B) MMP2 has AP2 and SP1 binding sites, which are found in promoter regions of constitutively expressed enzymes. MMP1, MMP3, MMP7, and MMP9 have AP1 binding sites, which are responsive to oncogenes and cytokines, consistent with their expression during inflammation. NF-κB is another inflammatory binding site. *AP*, activator protein; *C/EBP*, CCAAT/enhancer-binding protein; *CIZ*, CAS-interacting zinc-finger protein; *CREB*, cyclic AMP response-element binding protein; *EGR*, early growth response site; *FN*, fibronectin. *HP*, haemopexin; *ISE*, immortalization-sensitive element; *KRE*, keratinocyte differentiation-factor responsive element; *MMP*, matrix metalloproteinase; *MT*, membrane type; *NF-κB*, nuclear factor-κB; *PEA*, polyomavirus enhancer-A binding-protein; *Pre-Pro*, prepropeptide; *RCE*, retinoblastoma control element; *SP*, specificity protein; *SPRE*, stromelysin-1 platelet-derived growth factor β responsive element; *STAT*, signal transducer and activator of transcription; *TBS*, translocation-ETS-leukemia binding site; *TCF*, T-cell factor site; *TIE*, transforming growth factor β inhibitory element; *TMD*, transmembrane domain; *ZBP*, zinc-binding protein. *Adapted from Overall and Lopez-Otin, 5 with permission from Macmillan Publishers Ltd.*

vessels. When MMP-2 is activated it concentrates its actions close to the basal lamina. On the contrary, the active MMP-9 is released into the extracellular space where it is able to cause greater damage (Fig. 5.4) (Candelario-Jalil et al., 2009).

The pericytes are cells embedded in the basal lamina. Acting as macrophages, they release MMP-3, which is important in opening of the blood–brain barrier (Yang et al., 2018). Pericytes have both smooth muscle and macrophage functions. Because of their proximity to the endothelial cells, they are important contributing to both injury and repair (Dore-Duffy, 2008). Pericytes act like both macrophages to promote inflammation and smooth muscle cells to regulate capillary blood flow (Fig. 5.5).

Both pericytes and endothelial cells share a basement membrane where a variety of intercellular connections are made (Winkler et al., 2011). Pericytes have the same lineage to other cell types, including smooth muscle cells, neural cells, NG2 glia, muscle fibers, adipocytes, as well as fibroblasts and

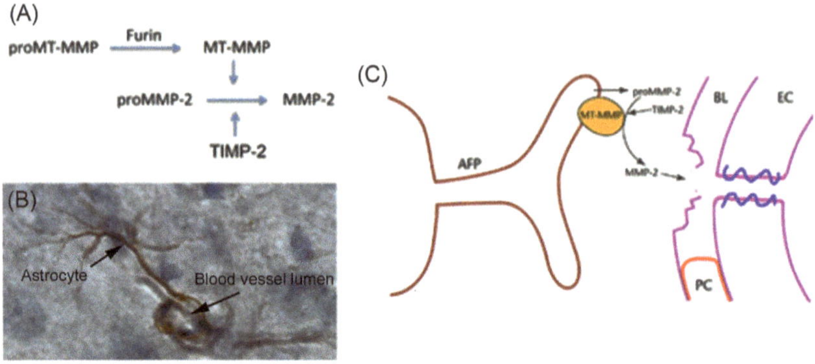

FIGURE 5.4 **Action of the constitutive enzyme, MMP-2, under normal conditions.** (A) Mechanism of activation of MMP-2. Furin activates proMT-MMP. A trimolecular complex composed of MT-MMP, proMMP-2, and tissue inhibitor to MMP-2 (TIMP-2) leads to the activated form of MMP-2. (B) A section of brain tissue from a normal rat immunostained with MMP-2, showing the localization to the astrocyte with the foot process wrapped around the blood vessel. The tissue is counterstained with Nissl stain. (C) A schematic drawing of the localization of the MT-MMP on the astrocyte foot process; the activation of MMP-2 is done close to the surface of the blood vessel preventing widespread damage. The basal lamina is broken down by MMP-2 and tight junction of the EC is disrupted. PC are not involved in this process. *EC*, Endothelial cell; *PC*, pericytes.

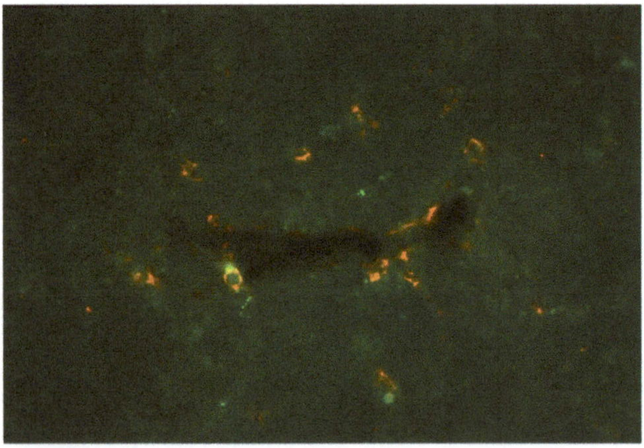

FIGURE 5.5 **Pericytes containing MMP-3 in pericytes.** An immunostained tissue from rat brain showing a blood vessel (*black area* in center) surrounded by pericytes. The pericytes are stained with a green fluorescent label and the orange label is MMP-3.

other mesenchymal stem cells. Pericytes play a crucial role in the formation and functionality of the blood–brain barrier. Pericytes are responsible for tight junction formation and vesicle trafficking in endothelial cells. Mice lacking pericytes are unable to maintain normal blood–brain barrier permeability (Bell et al., 2010). Aging causes a loss of pericyte regenerative capacity. In animal models with lower pericyte coverage, trafficking of molecules across endothelial cells occurs at a higher frequency, allowing proteins into the brain that would normally be excluded. Loss or dysfunction of pericytes is also theorized to contribute to neurodegenerative diseases such as Alzheimer's and Parkinson's through breakdown of the blood–brain barrier.

Angiogenesis

Angiogenesis is an important process during development and recovery from injury. The process of angiogenesis involves sprouting from existing blood vessels; after ischemia or during hypoxia, there is stimulation of angiogenesis, which participates in recovery. However, vascular permeability is increased in the newly formed vessels; the combination of beneficial and detrimental effects has made the role of angiogenesis in brain injury difficult to assess. In acute ischemia, the stimulus for the new vessels is clearly a part of the injury process in the early stages. Chronic forms of vascular injury do not have the acute phase, but the injury process evolves more slowly. When the cerebral blood vessels are changed over time by hypertension and other vascular risk factors, there is a mechanical impairment in the flow of blood and the tissues become hypoxic. As opposed to acute loss of blood flow in stroke where loss of blood flow is abrupt, when the hypoxia is intermittent, it paradoxically causes a greater impact on HIF-1α induction, maintaining higher levels for a longer time; sleep apnea is a example of intermittent hypoxia (Semenza & Prabhakar, 2018). Reduced cerebral blood flow with hypoxic hypoperfusion is a stimulus for angiogenesis in both AD and VCID, which has been shown in pathological studies (Biron et al., 2011; Kalaria et al., 1998; Tarkowski et al., 2002). Hypoxia is a strong stimulus for the induction of angiogenic factors: VEGF is formed under hypoxic conditions, and provides a major stimulus for angiogenesis. Blood vessels contain receptors for VEGF, such as FMS related receptor tyrosine kinase 1 (Flt-1) and Angiopoietin-1 receptor (Tie-2).

Angiogenesis is a multi-step process that requires a major contribution from proteases: degradation of the extracellular matrix around the blood vessels is an early step in the process. Stromelysins are matrix-degrading proteases that separate pericytes from the outer surface of blood vessels and remove extracellular matrix to facilitate the breakdown of the tissue around the vessel, which is necessary for sprouting of new vessels. Most of the information on the interaction role of the proteases in angiogenesis is derived from studies in cancer where growth of cancer cells requires a parallel growth of the cancer vasculature. Agents that block angiogenesis have been extensively explored in the treatment of cancer, such as agents that block VEFG (Folkman, 1971). Much less is known about the role of angiogenesis in brain; there is evidence to suggest that angiogenesis is beneficial in brain and blocking it could have negative effects. Hypoxia is a strong stimulus for angiogenesis in acute stroke. Presumably in brain similarly to cancer cells this cooperation of the proteases in the preparation for the vessel growth is a critical factor (Rundhaug, 2005). In an animal model of chronic hypertension using the spontaneously hypertensive stroke prone rat, there is evidence of angiogenesis related to protease production, particularly MMP-3 and expression of HIF-1α and VEGF (Yang et al., 2018). In both AD and VCID there is evidence that the cerebral blood flow is reduced, probably by different mechanisms. Hypoxia results from reduced cerebral blood flow during hypoperfusion, which induces the oxygen sensor molecule, HIF-1α, initiating a cassette of molecular events to increase red blood cells, promote anaerobic glycolysis, and releases the protein convertase, furin, which activates metalloproteinases. Both VEGF-C and its receptor Flt-1 were elevated in the CSF (Erhardt et al., 2021). Flt-1 is reported to be a signaling factor for microglia, which indicates that Flt-1 both acts on the blood vessels and on the microglia.

There is evidence that VEGF is protective in AD patients. A higher VEGF level was associated with larger baseline hippocampal volume, less hippocampal atrophy over time, and less cognitive decline over time; neuroprotective effect of VEGF appeared strongest in the presence of enhanced AD biomarkers, suggesting that angiogenic factors may be particularly important in those individuals showing early hallmarks of the AD cascade (Hohman et al., 2015).

VEGF has several forms each with unique functions: in brain the major forms are VEGF-A, VEGF-C and VEGF-D. Placental growth factor (PLGF) is a member of the angiogenesis family and is particularly related to inflammation. The receptors for the various forms of VEGF have been identified and linked to angiogenesis. Two distinct receptors for VEGF, the tyrosine kinase receptors, VEGFR-1 (Flt-1) and VEGFR-2 (Flk-1/KDR), have been identified (Fig. 5.6). Transfection studies demonstrated biological activities for the Flk-1/KDR-, but not for the Flt-1-receptor, which led to the hypothesis that Flt-1 is a decoy receptor for VEGF. However, Flt-1 is biologically active in nonendothelial cells, namely monocytes, which exclusively express this receptor. In addition, the Flt-1 ligand PLGF, which is unable to bind and activate Flk-1/KDR, elicits activities in both monocytes and endothelial cells. The two endothelial VEGF receptors belong to the receptor tyrosine kinases: the fms-like tyrosine kinase Flt-1 (human and murine), and the fetal liver kinase, Flk-1 (derived from the mouse) or its human homolog kinase insert domain-containing receptor (KDR). Flt-1 but not Flk-1/KDR is recognized by the secreted PLGF shows strong homology to VEGF. Tissue culture studies with PLGF revealed a mitogenic activity in endothelial cells, although considerably less than that observed with VEGF. This may be explained by the selective affinity of PLGF for one of the VEGF receptors. The first

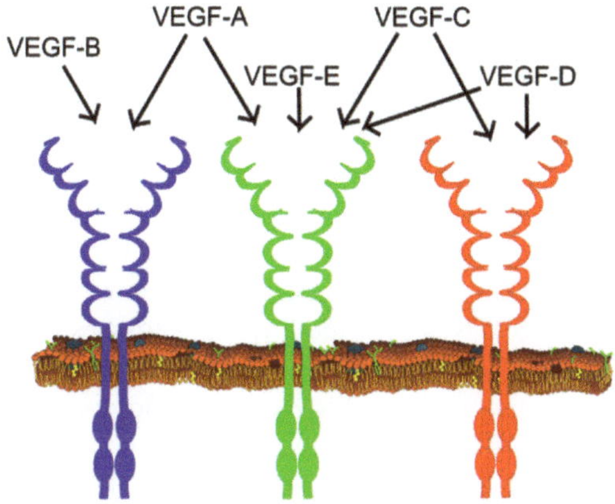

FIGURE 5.6 **VEGF types and receptors.** Google figure which can be used by listing the name of the person creating the figure. Illustration of the two major VEGF receptors Flt-1 and Flk-1/KDR. The forms of VEGF that activate these receptors are shown.*VEGF*,Vascular endothelial growth factor. *Used with permission of Mikael Häggström.*

evidence for PLGF as a specific ligand for Flt-1 came from competitive binding studies performed with the extracellular domains from either KDR/Flk-1 or Flt-1. Subsequently, studies with isolated peripheral blood monocytes and genetically engineered cells which express either one of the two VEGF receptors, confirmed that PLGF is able to bind to Flt-1 but not to KDR/Flk-1 on living cell (Clauss, 1998).

Cytokines

Cytokines are small proteins ($\sim$5–20 kDa) important in cell signaling that are involved in autocrine, paracrine, and endocrine signaling as immunomodulating agents. They include chemokines, interferons, interleukins, lymphokines, and tumor necrosis factors (TNF), and are produced by immune cells like macrophages, B lymphocytes, T lymphocytes, and mast cells, as well as endothelial cells, fibroblasts, and various stromal cells. Acting through cell surface receptors, they modulate the balance between humoral and cell-based immune responses. A classification that proves more useful in clinical and experimental practice outside of structural biology divides immunological cytokines into those that enhance cellular immune responses, type 1 (TNFα, IFN-γ, etc.), and those that favor antibody responses or type 2 (TGF-β, IL-4, IL-10, IL-13, etc.). A key focus of interest has been that cytokines in one of these two subsets tend to inhibit the effects of those in the other. Dysregulation of this tendency is under intensive study for its possible role in the pathogenesis of autoimmune disorders. Several inflammatory cytokines are induced by oxidative stress. The fact that cytokines themselves trigger the release of other cytokine, and also lead to increased oxidative stress makes them important in chronic inflammation. Cytokines also play a role in anti-inflammatory pathways and are a possible therapeutic treatment for pathological pain from inflammation or peripheral nerve injury.

Type-1 cytokines are cytokines produced by Th1 T-helper cells while Type-2 cytokines are those produced by Th2 T-helper cells. Type-1 cytokines include interleukin-2 (IL2), interferon-gamma (IFN-γ), IL-12 and tumor necrosis factor-β (TNF-β), while Type 2 cytokines include IL-4, IL-5, IL-6, IL-10, and IL-13. A classification that proves more useful in clinical and experimental practice outside of structural biology divides immunological cytokines into those that enhance cellular immune responses, type 1 (TNFα, IFN-γ, etc.), and type 2 (TGF-β, IL-4, IL-10, IL-13, etc.), which favor antibody responses. A key focus of interest has been that cytokines in one of these two subsets tend to inhibit the effects of those in the other. Dysregulation is thought to be important in the pathogenesis of autoimmune disorders. Several inflammatory cytokines are induced by oxidative stress. Study of the role of cytokines in brain inflammation has been hindered since they have both pro-inflammatory and anti-inflammatory actions in the brain.

Il-2 amplifies regulatory T-cells (T_{reg}) that are linked to chemokines, CCL1 and CCL20, which suppress astrocytosis and contribute to repair (Ito et al., 2019). IL-13 controls microglia function by converting M1/M2 microglia into anti-inflammatory M2 in traumatic brain injury (Miao et al., 2020). A recent study found that resilient patients with AD proteins had different patterns of cytokine expression; increased levels of IL-13 in the brains of resilient forms of AD. They linked these differences to reduced glial activation, increased neuronal survival and preserved cognition in more resilient cases (Barroeta-Espar et al., 2019). In vivo expansion of regulatory T cells with IL-2/IL-2 antibody complex protects against ischemic attacks. However, the low frequency of T_{regs} restricts their clinical utility. Systemic administration of the IL-2/IL-2 antibody complex (IL-2/IL-2Ab) robustly and selectively expanded the number of T_{regs} after stroke. Furthermore, IL-2/IL-2Ab not only elevated T_{reg} numbers, but also boosted their functions and enhanced the expression of CD39 and CD73. CD73-deficient mice showed its importance in the protective action of IL-2/IL-2Ab-stimulated Tregs in mice with strokes. These results shed light on IL-2/IL-2Ab as a clinically feasible immune therapy to boost endogenous Treg responses and ameliorate ischemic brain injury (Zhang et al., 2018).

Il-10 has a protective role in damping down the inflammatory response (Lobo-Silva et al., 2016). On the surface of microglia, there are many pattern recognition receptors (PRRs) containing scavenger receptors and toll-like receptors (TLRs). Danger-associated molecular patterns (DAMP) including misfolded protein, mislocalized nucleic acids and peptides, and pathogen-associated molecular patterns (PAMP) can bind to PRRs and subsequently initiate a series of immune responses, ultimately inducing the production of proinflammatory cytokines by microglia. It is well known that microglia has two phenotypes including proinflammatory (M1) form and anti-inflammatory (M2) form. The classic M1 phenotype is characterized by the delivery of proinflammatory cytokines including tumor necrosis factor-α (TNF-α), interleukin (IL)-1, -6, -12, and -18, accompanied with impaired phagocytic function while the alternative is marked by the delivery of anti-inflammatory cytokines containing IL-4, IL10 and transforming growth factor-β (TGF-β), accompanied with increased phagocytic function (Wang et al., 2018). Taken together, the available data indicates that IL-10 plays a critical role in limiting CNS inflammation in a similar manner to that seen at peripheral sites by altering the ability of resident glia and infiltrating leukocytes to respond to activating stimuli, and by reducing the production of inflammatory mediators by these cells (Burmeister & Marriott, 2018).

M2 macrophages facilitate resolution of inflammation through anti-inflammatory factors (e.g., IL-10, IL-13, TGF-β), VEGF, EGF, Arg1 to deactivate pro-inflammatory cell phenotypes and reestablish homeostasis. This includes production of IL-10 to down regulate inflammatory cells, extracellular matrix protecting proteins like Ym1/2, ornithine, polyamides for wound repair and higher levels of receptors associated with phagocytosis. IL-10 induces STAT3 and downstream genes including *Il10*, *Tgfb1*, macrophage mannose receptor *Mrc1*. Based upon peripheral macrophage nomenclature, M1 and M2 polarization states of microglia have been proposed as a framework to evaluate the heterogeneity of responses (Mosser & Edwards, 2008) with recent evaluations suggesting a framework focused on the inducing stimuli (Martinez & Gordon, 2014). This is based upon papers adapted from Martinez and Gordon (2014). Under the classic M1 state, exposure to LPS and/or IFN-γ stimulates TLR4 or IFN-γ receptors 1 and 2, respectively, leading to activation of transcription factors NF-κβ and STAT1 and increased expression of CD86 and MHC-II. The increase in iNOS produces a burst of ROS and reactive nitrogen species (RNS) and the release of pro-inflammatory cytokines, such as IL-1α, IL-1β, TNF-α, IL-6, IL-12, IL-23, the chemokines CCL2 and CCL20, and the receptor CCR2, macrophage receptor with collagenous structure (MARCO) and COX2. M2 states have been proposed following various stimulatory factors. For example, upon stimulation with IL-4/IL-13, binding of the IL-4 receptor (IL-4R) initiates activation of STAT6, shifting the cells towards an anti-inflammatory phenotype with an increase in Arg-1, expression of CD206 and mannose receptor (Mr), and release of anti-inflammatory factors IL-4, IL-6, IL-10, IL-13, IL-1RA, FIZZ1, and PPARγ. Exposure of the cells to IL-10 activates STAT6 via stimulation of IL-10 receptors 1 and 2. This serves to shift the cells to a primary immunosuppressive state with an expression of CD206 and the release of IL-10, TGF-β, FIZZ1 and PPARγ (Orihuela et al., 2016).

In summary, IL-2, IL-10, and IL-13 have properties that reduce astrogliosis. IL-2 acts through the T_{reg} lymphocytes that normally block gliosis by inducing the production of IL-10, which is the anti-inflammatory cytokine involved in the production. Another mechanism is used by IL-13 to dampen inflammatory gliosis: it converts the M1 microglia cell, which is proinflammatory to the anti-inflammatory M2 phenotype. These conclusions on the action of the anti-inflammatory cytokines are based primarily on models of disease other than those related to dementia; it will be interesting to determine their role in animal models of dementia.

References

Barroeta-Espar, I., Weinstock, L. D., Perez-Nievas, B. G., Meltzer, A. C., Siao Tick Chong, M., Amaral, A. C., Murray, M. E., Moulder, K. L., Morris, J. C., Cairns, N. J., Parisi, J. E., Lowe, V. J., Petersen, R. C., Kofler, J., Ikonomovic, M. D., López, O., Klunk, W. E., Mayeux, R. P., Frosch, M. P., & Gomez-Isla, T. (2019). Distinct cytokine profiles in human brains resilient to Alzheimer's pathology. *Neurobiology of Disease*, *121*, 327–337. Available from https://doi.org/10.1016/j.nbd.2018.10.009.

Bell, R. D., Winkler, E. A., Sagare, A. P., Singh, I., LaRue, B., Deane, R., & Zlokovic, B. V. (2010). Pericytes control key neurovascular functions and neuronal phenotype in the adult brain and during brain aging. *Neuron*, *68*(3), 409–427. Available from https://doi.org/10.1016/j.neuron.2010.09.043.

Biron, K. E., Dickstein, D. L., Gopaul, R., & Jefferies, W. A. (2011). Amyloid triggers extensive cerebral angiogenesis causing blood brain barrier permeability and hypervascularity in Alzheimer's disease. *PLoS One*, *6*(8). Available from https://doi.org/10.1371/journal.pone.0023789.

Brinckerhoff, C. E., & Matrisian, L. M. (2002). Matrix metalloproteinases: A tail of a frog that became a prince. *Nature Reviews. Molecular Cell Biology*, *3*(3), 207–214.

Burmeister, A. R., & Marriott, I. (2018). The interleukin-10 family of cytokines and their role in the CNS. *Frontiers in Cellular Neuroscience*, *12*, 458. Available from https://doi.org/10.3389/fncel.2018.00458.

Candelario-Jalil, E., Yang, Y., & Rosenberg, G. A. (2009). Diverse roles of matrix metalloproteinases and tissue inhibitors of metalloproteinases in neuroinflammation and cerebral ischemia. *Neuroscience*, *158*(3), 983–994. Available from https://doi.org/10.1016/j.neuroscience.2008.06.025.

Clauss, M. (1998). Functions of the VEGF receptor-1 (FLT-1) in the vasculature. *Trends in Cardiovascular Medicine*, *8*(6), 241–245.

Cserr, H. F., Harling-Berg, C. J., & Knopf, P. M. (1992). Drainage of brain extracellular fluid into blood and deep cervical lymph and its immunological significance. *Brain Pathology (Zurich, Switzerland)*, *2*(4), 269–276.

Davson, H. (1976). Review lecture. The blood-brain barrier. *The Journal of Physiology*, *255*(1), 1–28.

Dore-Duffy, P. (2008). Pericytes: pluripotent cells of the blood brain barrier. *Current Pharmaceutical Design*, *14*(16), 1581–1593.

Erhardt, E. B., Adair, J. C., Knoefel, J. E., Caprihan, A., Prestopnik, J., Thompson, J., Hobson, S., Siegel, D., & Rosenberg, G. A. (2021). Inflammatory Biomarkers Aid in Diagnosis of Dementia. *Frontiers in Aging Neuroscience*, *13*, 717344. Available from https://doi.org/10.3389/fnagi.2021.717344.

Folkman, J. (1971). Tumor angiogenesis: therapeutic implications. *The New England Journal of Medicine*, *285*(21), 1182–1186.

Hawkins, B. T., & Davis, T. P. (2005). The blood-brain barrier/neurovascular unit in health and disease. *Pharmacological Reviews*, *57*(2), 173–185.

Hohman, T. J., Bell, S. P., & Jefferson, A. L. (2015). The role of vascular endothelial growth factor in neurodegeneration and cognitive decline: exploring interactions with biomarkers of Alzheimer disease. *JAMA Neurology*, *72*(5), 520–529. Available from https://doi.org/10.1001/jamaneurol.2014.4761.

Iadecola, C., & Nedergaard, M. (2007). Glial regulation of the cerebral microvasculature. *Nature Neuroscience*, *10*(11), 1369–1376.

Ito, M., Komai, K., Mise-Omata, S., Iizuka-Koga, M., Noguchi, Y., Kondo, T., Sakai, R., Matsuo, K., Nakayama, T., Yoshie, O., Nakatsukasa, H., Chikuma, S., Shichita, T., & Yoshimura, A. (2019). Brain regulatory T cells suppress astrogliosis and potentiate neurological recovery. *Nature*, *565*(7738), 246–250. Available from https://doi.org/10.1038/s41586-018-0824-5.

Kaelin, W. G., & Ratcliffe, P. J. (2008). Oxygen sensing by metazoans: the central role of the HIF hydroxylase pathway. *Molecular Cell*, *30*(4), 393–402. Available from https://doi.org/10.1016/j.molcel.2008.04.009.

Kaelin, W. G., Ratcliffe, P. J., & Semenza, G. L. (2016). Pathways for oxygen regulation and homeostasis: The 2016 Albert Lasker basic medical research award. *JAMA: the Journal of the American Medical Association*, *316*(12), 1252–1253. Available from https://doi.org/10.1001/jama.2016.12386.

Kalaria, R. N., Cohen, D. L., Premkumar, D. R. D., Nag, S., Lamanna, J. C., & Lust, W. D. (1998). Vascular endothelial growth factor in Alzheimer's disease and experimental cerebral ischemia. *Molecular Brain Research*, *62*(1), 101–105. Available from https://doi.org/10.1016/S0169-328X(98)00190-9.

King, L. S., Yasui, M., & Agre, P. (2000). Aquaporins in health and disease. *Molecular Medicine Today*, *6*(2), 60–65.

Liotta, L. A., & Stetler-Stevenson, W. G. (1990). Metalloproteinases and cancer invasion. *Seminars in Cancer Biology*, *1*(2), 99–106.

Lobo-Silva, D., Carriche, G. M., Castro, A. G., Roque, S., & Saraiva, M. (2016). Balancing the immune response in the brain: IL-10 and its regulation. *Journal of Neuroinflammation*, *13*(1), 297.

Louveau, A., Plog, B. A., Antila, S., Alitalo, K., Nedergaard, M., & Kipnis, J. (2017). Understanding the functions and relationships of the glymphatic system and meningeal lymphatics. *The Journal of Clinical Investigation*, *127*(9), 3210–3219. Available from https://doi.org/10.1172/JCI90603.

Martinez, F. O., & Gordon, S. (2014). The M1 and M2 paradigm of macrophage activation: time for reassessment. *F1000prime Rep*, *6*, 13. Available from https://doi.org/10.12703/P6-13.

Miao, W., Zhao, Y., Huang, Y., Chen, D., Luo, C., Su, W., & Gao, Y. (2020). IL-13 ameliorates neuroinflammation and promotes functional recovery after traumatic brain injury. *Journal of Immunology (Baltimore, Md.: 1950)*, *204*(6), 1486–1498. Available from https://doi.org/10.4049/jimmunol.1900909.

Mosser, D. M., & Edwards, J. P. (2008). Exploring the full spectrum of macrophage activation. *Nat Rev Immunol*, *8*(12), 958–969. Available from https://doi.org/10.1038/nri2448.

Orihuela, R., McPherson, C. A., & Harry, G. J. (2016). Microglial M1/M2 polarization and metabolic states. *British Journal of Pharmacology*, *173*(4), 649–665. Available from https://doi.org/10.1111/bph.13139.

Punchard, N.A., Whelan, C.J., & Adcock, I. (2004). Journal of Inflammation (London, England), 1(1), 1.

Reese, T. S., & Karnovsky, M. J. (1967). Fine structural localization of a blood-brain barrier to exogenous peroxidase. *The Journal of Cell Biology*, *34*(1), 207–217.

Reeves, B. C., Karimy, J. K., Kundishora, A. J., Mestre, H., Cerci, H. M., Matouk, C., Alper, S. L., Lundgaard, I., Nedergaard, M., & Kahle, K. T. (2020). Glymphatic system impairment in Alzheimer's disease and idiopathic normal pressure hydrocephalus. *Trends in Molecular Medicine*, *26*(3), 285–295. Available from https://doi.org/10.1016/j.molmed.2019.11.008.

Rosenberg, G. A. (2009). Matrix metalloproteinases and their multiple roles in neurodegenerative diseases. *The Lancet Neurology*, *8*(2), 205–216. Available from https://doi.org/10.1016/S1474-4422(09)70016-X.

Rosenberg, G. A., Kyner, W. T., & Estrada, E. (1980). Bulk flow of brain interstitial fluid under normal and hyperosmolar conditions. *The American Journal of Physiology*, *238*(1), F42–F49.

Rundhaug, J. E. (2005). Matrix metalloproteinases and angiogenesis. *Journal of Cellular and Molecular Medicine*, *9*(2), 267–285. Available from https://doi.org/10.1111/j.1582-4934.2005.tb00355.x.

Semenza, G. L., & Prabhakar, N. R. (2018). The role of hypoxia-inducible factors in carotid body (patho) physiology. *The Journal of Physiology*, *596*(15), 2977–2983. Available from https://doi.org/10.1113/JP275696.

Tarkowski, E., Issa, R., Sjögren, M., Wallin, A., Blennow, K., Tarkowski, A., & Kumar, P. (2002). Increased intrathecal levels of the angiogenic factors VEGF and TGF-β in Alzheimer's disease and vascular dementia. *Neurobiology of Aging*, *23*(2), 237–243. Available from https://doi.org/10.1016/S0197-4580(01)00285-8.

Wang, M.-M., Miao, D., Cao, X.-P., Tan, L., & Tan, L. (2018). Innate immune activation in Alzheimer's disease. *Annals of Translational Medicine*, *6*(10), 177. Available from https://doi.org/10.21037/atm.2018.04.20.

Weed, L. H. (1914). Studies on Cerebro-Spinal Fluid. No. III: The pathways of escape from the Subarachnoid Spaces with particular reference to the Arachnoid Villi. *The Journal of Medical Research*, *31*(1), 51–91.

Winkler, E. A., Bell, R. D., & Zlokovic, B. V. (2011). Central nervous system pericytes in health and disease. *Nature Neuroscience*, *14*(11), 1398–1405. Available from https://doi.org/10.1038/nn.2946.

Yang, Y., Kimura-Ohba, S., Thompson, J. F., Salayandia, V. M., Cossé, M., Raz, L., Jalal, F. Y., & Rosenberg, G. A. (2018). Vascular tight junction disruption and angiogenesis in spontaneously hypertensive rat with neuroinflammatory white matter injury. *Neurobiology of Disease*, *114*, 95–110. Available from https://doi.org/10.1016/j.nbd.2018.02.012.

Zhang, H., Xia, Y., Ye, Q., Yu, F., Zhu, W., Li, P., Wei, Z., Yang, Y., Shi, Y., Thomson, A. W., Chen, J., & Hu, X. (2018). Expansion of regulatory T cells with IL-2/IL-2 antibody complex protects against transient ischemic stroke. *The Journal of Neuroscience: The Official Journal of the Society for Neuroscience*, *38*(47), 10168–10179. Available from https://doi.org/10.1523/JNEUROSCI.3411-17.2018.

Chapter 6

Animal models for vascular cognitive impairment

Introduction

Disease models are used for studies of pathology, identification of molecular and cellular mechanisms, pre-clinical testing of potential treatments, and confirmation of clinical biomarkers. The models are either mimics of vascular cognitive impairment (VCI)-related brain lesions, such as white matter hypoperfusion injury, focal ischemia, cerebral amyloid angiopathy, or they reproduce VCI risk factors, such as old age, hypertension, hyperhomocysteinemia, high-salt/high-fat diet and diabetes. Finally, there are animal models that reproduce genetic causes of VCI. None of the current models fully reflect the human disease. The most useful current models are based on rodents, which have little white matter, while larger animals with greater amounts of white matter are more difficult to study. There is a need for further research on translational models (Hainsworth et al., 2017).

Vascular disease in itself is a major cause of dementia and contributes to other causes. Vascular cognitive impairment and dementia (VCID) is the term used to describe all the forms of dementia related to vascular disease: VCID is a spectrum of clinical conditions, ranging from large vessel disease with post-stroke cognitive impairment to small vessel disease (SVD). There are genetic forms of small vessel arteriopathy related to specific gene mutations, such as CADASIL, CARASIL, and COL4A1/4A2. SVD is the most common pathology underlying VCID, making it the most important. Modeling this diverse collection of pathology is challenging and prevents any single model from representing the entire group.

The majority of studies in VCID are done with rodents since small animals are easier to handle, cheaper, and have similar molecular pathways as humans. Genetic studies are generally done with mice due to the ease of manipulating the genes with molecular biology methods. Neuroimaging of white matter damage is improved with larger animals, such as pigs, and rhesus and cromalygous monkeys.

Neuroinflammation in Vascular Dementia. DOI: https://doi.org/10.1016/B978-0-12-823455-6.00004-3

Bilateral carotid artery occlusion in rat

Rodent animal models are the most commonly used in the study of VCI. The animal models need to demonstrate the pathophysiology. Although large animals have several advantages, they increase the complexity of the experiments and are, therefore used infrequently. For this reason, rodents are the most commonly used laboratory animals, with other species used less frequently and only in special circumstances, such as the use of nonhuman primates for advanced drug testing prior to use in humans. Advantages of the rat and mouse models are the ease of handling, lower cost, the ability to perform high field imaging at 4 Tesla and above, and the large number of reagents available for immunohistochemistry and biochemical analysis. Mice have the added advantage of having a large number of genetically manipulated animals that allow for the study of specific disease mechanisms. The majority of the studies on vascular causes of dementia are done in rats, while the development of newer models in mice have increased the number of mouse studies. An important benefit of mouse models is the ability to manipulate genes that makes the mouse the species of choice for studies in the diseases associated with dementia, such as Alzheimer's disease.

The first model developed to study VCID was the bilateral carotid artery occlusion (BCAO) in the normotensive rat (Wakita et al., 1994). This model produced white matter demyelination with oligodendrocyte (Ol) death; behavioral studies showed deficits compatible with hippocampal cell damage. A number of drugs were tested in this model with many showing therapeutic benefit. The drawback was the drop in cerebral blood flow (CBF) after occlusion of the carotids that lead to a period of acute ischemia that was not characteristic of the chronic progression of the human disease. A newer model in rats and mice that avoids the sudden fall in blood flow in the complete occlusion model is the use of an ameroid constrictor (AC) that gradually occludes the carotids, producing an injury that is closer to the human condition (Kitamura et al., 2012).

When both carotids are occluded and the CBF is restricted, there is hypoxic hypoperfusion. The hypoxia results in opening of the blood–brain barrier (BBB) by matrix metalloproteinases (MMPs) that are expressed by the reduction in oxygen. An increase in MMP-2 reported to occur, contributing to the disruption of the BBB (Ihara et al., 2001; Nakaji et al., 2006). Further studies showed an increase in MMP-3, which was also implicated in demyelination in a mouse model (D'Souza et al., 2002). The relationship of hypoxic hypoperfusion by BCAO to disruption of the BBB was studied by the combination of MRI and immunohistochemistry. MRI showed regions of increased apparent diffusion coefficients that correlated with extravasation of Evans blue, a marker of BBB opening that binds to albumin (Sood et al., 2009).

Hypoxic hypoperfusion in mice

Cerebral blood constriction can be done by very small coils placed around carotid arteries

The early attempts to use rodents as models of VCID that were done in rats with total occlusion of the carotid arteries resulted in temporary acute ischemia with gradual restoration of the CBF, but at a reduced level. These animals had normotension, which limited the usefulness of the model since there was no long-term damage to the blood vessels as seen in humans. Attempts to repeat these studies in animals with chronic hypertension caused the animals to die. Occlusion of both carotid arteries in a normotensive rat mimicked the pathophysiology VCI by causing damage to the BBB and myelin loss in the white matter through the action of MMP-2 (Candelario-Jalil et al., 2009). In addition to BBB opening, there is atrophy of the hippocampal neurons. Activated microglia in white matter maximally produced MMP-2 on day 3 of chronic cerebral hypoperfusion, preceding the rarefaction of the white matter. White matter is highly susceptible to chronic cerebral hypoperfusion, which also affects the gray matter; in the white matter there is injury to both axons and myelin. At the onset of the carotid occlusion, there is a period of interference with CBF, which causes acute cerebral ischemia. This fall in blood flow with ischemia is a major drawback of the BCAO. Another drawback is the damage to the visual system, which interferes with the cognitive testing. The BCAO model has been used extensively because of the ability to image the animals with MRI and to correlate the imaging findings with immunohistochemistry and biochemical studies. Another advantage is the large number of cognitive tests that have been standardized, particularly the Morris water maze (WM) that has become one of the standard tests of injury in the memory system (Wakita et al., 1994, 2002).

A more recent rodent model has been developed to overcome some of the drawbacks of the BCAO model. To produce a less severe impact on the CBF, the new model involves stenosis of the carotid arteries by metal coils of very small diameter. Many of the studies done in the BCAO model in the rat have been repeated in this new model with similar results, making this an excellent advance. This new model is produced by coils wrapped around the carotid arteries to limit blood flow without actual occlusion, avoiding the drop in blood flow that resulted in ischemic injury in the BCAO model. In addition to the avoidance of the fall in blood flow, the bilateral carotid artery stenosis model can be used to study genetically manipulated mice (Hajdu et al., 1991; Shibata et al., 2004).

The need to study animals with genetic modifications lead to the development of VCID models in mice. The first models in mice involved bilateral carotid artery stenosis (BCAS) employing the coils used in the rat studies;

tiny metal coils with very small diameters were wrapped around both common carotid arteries to induce cerebral hypoperfusion. These technically challenging models produced white matter lesions and memory impairment in mice similar to that seen in rats with either occlusion or stenosis. Although there was a limited amount of white matter in the mouse, high-field MRI showed the relationship of the white matter damage to memory impairment. There was a significant effect of BCAS on perceptual learning in the novel object recognition test and on number of errors and latency to platform in the radial arm water maze (RAWM). MRI analysis revealed a significant effect of BCAS on diffusion tensor imaging parameters in white matter regions; significantly lower fractional anisotropy (FA) values were found in the corpus callosum and anterior commissure and significantly higher mean diffusivity values in the internal capsule. Focusing on the corpus callosum, there was a correlation between FA and number of errors on the RAWM test, which were partly mediated by effects on white matter integrity. Immunofluorescence studies demonstrated significantly higher microglia cell density and soma size in the corpus callosum of BCAS mice compared to controls, and these parameters were correlated with the imaging data. The results of this study indicate that cognitive deficits induced by cerebral hypoperfusion due to BCAS result in part from microglia activation and disruption of white matter integrity, supporting the validity of this unique model of vascular dementia (Ben-Ari et al., 2019). Overall, the results of this study indicate that cerebral hypoperfusion due to BCAS micro coils is associated with cognitive deficits that are related to microglial activation and subsequent disruption of white matter integrity. A drawback to the use of coils is a high rate of mortality in the mice.

Ameroid constrictor bilateral carotid artery stenosis model

The BCAS model in mice is a highly promising model of chronic cerebral hypoperfusion; it exhibits characteristic features of human hypoperfusion in VCI, such as WM rarefaction and gliosis, and working memory impairment. However, this model has an inherent limitation in that CBF drops sharply immediately after BCAS surgery and then gradually recovers over 1 month, thus failing to reliably replicate "chronic" cerebral hypoperfusion apparent in VCI (Shibata et al., 2004). The contribution of the preceding acute phase to the neuropathological and behavioral consequences is therefore an ongoing concern. To overcome the limitations of the microcoil method of BCAS, another model was established that eliminated the acute phase: CBF gradually was decreased to the level found in the chronic phase by using a device called the AC, which predictably achieves gradual narrowing of the common carotid arteries.

In one study with the constrictor, Adult C57BL/6 J male mice were subjected to gradual common carotid artery stenosis (GCAS) surgery by using an ameroid constricting device with an inner diameter of 0.75 mm. The common carotid arteries narrowed gradually after gradual expansion of the ACs over 28 days after GCAS, with subsequent 79.3% area stenosis as a result of smooth muscle cell proliferation and macrophage infiltration in the tunica intima. The 28 day survival rate was 91%, which was much improved over the mortality in the animals subjected to carotid stenosis with a steel coil. Arterial spin labeling demonstrated gradual and continuous reduction of cortical and subcortical CBF (ratio to the preoperative value) to 54.6% and 51.5%, respectively, over 28 days. However, magnetic resonance angiography showed collateral flow signals in the leptomeningeal artery. Rarefaction and proliferation of astrocytes and microglia, with loss of Ols, were found in the white matter at 32 days. Hippocampal neuronal loss was observed in only 25% of GCAS mice, consistent with lack of abnormalities in the Morris WM test. The rotarod test showed motor impairment, and the Y-maze test showed working memory deficits. The GCAS model successfully generated gradual and continuous CBF reduction over 28 days, with replication of key histological, radiological, and behavioral features associated with cerebral hypoperfusion leading to VCI (Hattori et al., 2016).

Spontaneously hypertensive rats

The spontaneously hypertensive rat (SHR) is a laboratory rat which is an animal model of essential (or primary) hypertension, used to study cardiovascular disease and stroke. Measured by the number of published works, it is the most studied model of hypertension. The SHR strain was obtained during the 1960s by Okamoto and colleagues, who started breeding Wistar-Kyoto (WKY) rats with high blood pressure (Okamoto & Aoki, 1963). Stroke prone SHR (SHRSP) is a further development of SHR that has even higher blood pressure than SHR and increased mortality from stroke (Yamori et al., 1976).

The SHRSP showed edema in the gray matter associated with disruption of the BBB. There was extensive spread of antifibrinogen-positive material within the white matter that seemed to originate mainly from the chronic leakage sites in the gray matter. Increased number of large astrocytes were seen within the leakage sites and along the spreading pathways for the edema constituents. The white matter showed a rarefied texture with widely dispersed nerve fiber tracts, volume expansion, and occasional cyst formation. The results indicate a crucial pathophysiologic role for the egress, spread, and accumulation of vasogenic edema in the development of the cerebrovascular lesions in SHRSP (Fredriksson et al., 1988). Multifocal BBB opening with massive leakage of plasma constituents rich in fibrinogen-fibrin-related antigen occurred in SHRSP with a blood pressure above 210–220 mm Hg. BBB-leakage sites were found in the cerebral cortex and the basal ganglia, most frequently in the arterial border zones. The perivascular tissue spaces were dilated within the BBB-leakage sites, in particular around arterioles. Damaged endothelial and smooth muscle cells were replaced by fibrin-like material, multiple layers of basement membranes and bundles of collagen fibrils surrounded by proliferated fibroblasts (Fredriksson et al., 1988). These studies showed the vulnerability of the gray matter and the movement of vasogenic edema into the withe matter.

The onset of the hypertension in the SHRSP could be accelerated to shorten the time needed to study the disease process by feeding 7–8 week old rats with Japanese Permissive Diet (JPD), along with 1% salt in the water; 80% will develop cerebral hemorrhage at 16–18 weeks of age (Yamori et al., 1976). More importantly, after one month of the exaggerated diet, they developed characteristic lesions in the white matter similar to those seen in patients with the hypertensive form of VCID. When BCAO was done on SHRSP, it caused a high mortality rate, making this not useful as a model. However, it was found that unilateral carotid artery occlusion (UCAO) produced damage to the white matter with disruption of the BBB by an inflammatory reaction mediated by MMPs, allowing for study of white matter damage pathophysiology in an accelerated model (Jalal et al., 2012).

White matter injury in SHRSP

In a study of white matter injury in the permanent UCAO in SHRSP at 12 weeks of age, the animals were placed on a JPD and received 1% NaCl in drinking water. MRI, histology, biochemistry, and ELISA characterized white matter lesions, and cognitive impairment was tested by the Morris WM. White matter damage was observed at 4–5 weeks following UCAO/JPD. Immunoblotting showed marked reduction in myelin basic protein and up regulation of immature Ols. Mature Ols underwent caspase-3-mediated apoptosis. The Morris WM showed cognitive impairment. Abnormally appearing vessels were observed and surrounded by inflammatory-like cells. IgG extravasation and hemorrhage, indicating BBB disruption, was closely associated with MMP-9 expression. Lesions in white matter showed reactive astrocytes and activated microglia that expressed tumor necrosis factor-α (TNF-α). MMP-3 and MMP-9 were significantly increased, and MMP-2 was reduced in both astrocytes and Ol. These findings indicated that there was death by apoptosis of mature Ols with an increase in immature Ols. Increased MMP-3, MMP-9, and TNF-α were associated with myelin breakdown and BBB disruption: neuroinflammation was an important factor in white matter damage and Ol death (Jalal et al., 2012). White matter IgG content was observed within the area of external capsule, and corpus callosum of the occluded side (Fig. 6.1A). Extravasations of IgG were also demonstrated in the regions of external capsule, corpus callosum, and internal capsule of the nonoccluded side using ELISA. In addition, MMP-9 was detected in the area of IgG leakage (Fig. 6.1B). MMP-3 was not detected. Histological staining showed hemorrhage and perivascular infiltrates of inflammatory-like cells. There was loss of structure in the white matter surrounding the vessels (Fig. 6.1C).

Blood-brain barrier opening and matrix metalloproteinases in SHRSP

Extravasation of the IgG, which is normally in the blood, occurs when the BBB is disrupted; white matter IgG content was observed within the area of the external capsule, and corpus callosum of the occluded side. Extravasations of IgG were also demonstrated in the regions of the external capsule, corpus callosum, and internal capsule of the nonoccluded side using ELISA (Figs. 6.2C and Fig. 6.3A). In addition, MMP-9 was detected in the area of IgG leakage (Fig. 6.3B). MMP-3 was not detected. Histological staining showed hemorrhage and perivascular infiltrates of inflammatory-like cells. There was loss of structure in the white matter surrounding the vessels.

Immunoreactivity with glia fibrillary acidic protein and CC1, a marker for Ols, was observed. No colocalization was observed with the micoglia marker, OX-42, suggesting that astrocytes and Ols are important sources of MMP-3 and MMP-9. Confocal immunohistochemistry images exhibited cleaved caspase-3, a marker of apoptotic cell death, positive cells in the area of external capsule, colocalized with CC1, suggesting that these Ols were undergoing apoptosis (Fig. 6.3A). Western blot analysis demonstrated increased cleaved caspase-3 in the areas of the external capsule, corpus callosum, and internal capsule of both hemispheres (Fig. 6.4B and C). Sham-operated group showed little cleaved caspase-3 (Jalal et al., 2012).

Hypoxia in SHRSP

It is possible to monitor oxygen levels in vivo in small regions of brain with electron paramagnetic resonance (EPR). To investigate whether a state of chronic hypoxia exists in SHR/JPD, we relied on a minimally invasive EPR oximetry for the study of in vivo cortical pO_2 levels. A lithium phthalocyanine crystal, which is an EPR probe detecting pO_2 levels, was surgically implanted in the right dorsal lateral cortex brain region at 10 W of life in experimental and control groups. The probe location was confirmed in postmortem tissue slices, as indicated in Fig. 6.4A. A time course of EPR recordings in live rats for interstitial pO_2 levels from 10 to 15 W was performed (Fig. 6.4B). We observed a significant drop in pO_2 of SHRSP/JPD at 12 W ($P < 0.05$), which remained consistently low until 15 W (13 W: $P < 0.01$,

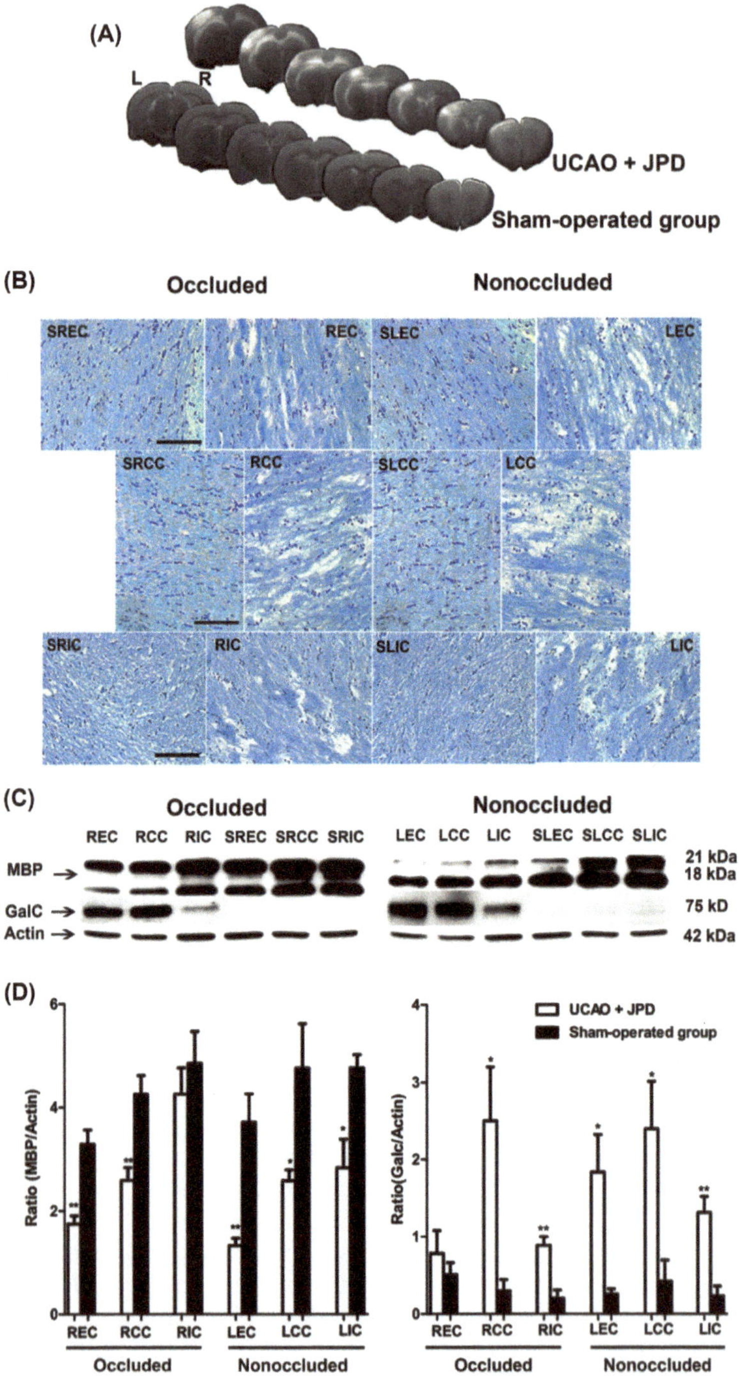
(A)
L
R
UCAO + JPD
Sham-operated group
(B)
Occluded
Nonoccluded
SREC
REC
SLEC
LEC
SRCC
RCC
SLCC
LCC
SRIC
RIC
SLIC
LIC
(C)
Occluded
Nonoccluded
REC RCC RIC SREC SRCC SRIC
LEC LCC LIC SLEC SLCC SLIC
MBP
GalC
Actin
21 kDa
18 kDa
75 kD
42 kDa
(D)
Ratio (MBP/Actin)
Ratio(Galc/Actin)
UCAO + JPD
Sham-operated group
REC RCC RIC LEC LCC LIC
Occluded
Nonoccluded

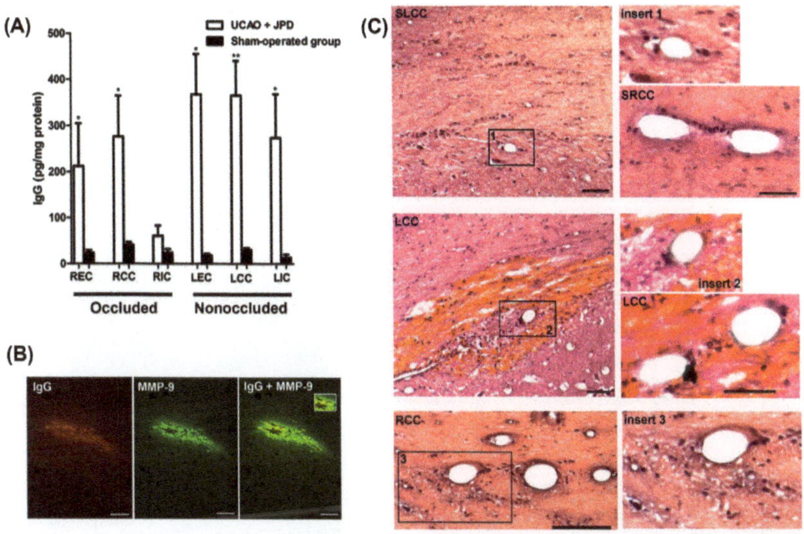

FIGURE 6.2 Blood–brain barrier opening in SHRSP blood–brain barrier damage following UCAO/JPD. (A) Quantification of IgG concentration by ELISA for both occluded (R) and non-occluded (L) sides of EC, CC, and IC. Values are expressed as mean ± S.E.M. (B) Immunohistochemistry analysis shows IgG leakage from vessels in the CC of UCAO/JPD group, IgG was co-localized with MMP-9 (*yellow*). (C) H&E-stained sections revealed hemorrhages surrounding vessels in the area of CC in UCAO/JPD group. No hemorrhage was seen in SCC (upper four panels). Perivascular infiltrate of inflammatory like cells in both occluded (R) and nonoccluded (L) sides of CC following UCAO/JPD (lower four panels). Well maintained blood vessels in sham-operated control. Scale bar, 50 μm (right panels) and 100 μm (left panels). * $P < 0.05$, ** $P < 0.01$ compared with corresponding control ($n = 5$/group). *CC*, Corpus callosum; *EC*, external capsule; *IC*, internal capsule; *JPD*, Japanese permissive diet; *SCC*, sham-operated group corpus callosum; *SHRSP*, stroke prone spontaneously hypertensive rat; *UCAO*, unilateral carotid artery occlusion.

FIGURE 6.1 Myelin loss in SHRSP with UCAO and up-regulation of immature Ols seen at 4–5 weeks following UCAO/JPD. (A) T2-weighted images obtained from UCAO/JPD and sham-operated groups demonstrate hyperintense areas on both occluded (R) and nonoccluded (L) sides. An infarct seen in nonoccluded cortex. (B) Klüver-Barrera staining showed no damage in SEC, SCC, and SIC in both hemispheres. UCAO/JPD rats had myelin and cell loss, as well as vacuolation, on both occluded and nonoccluded sides of EC, CC, and IC. (C) Representative blots of myelin basic protein (MBP) and galactosylceramidase gene (GalC) for occluded and nonoccluded sides of different areas of white matter. Actin was used as a loading control. (D) Western blot densitometric analysis of MBP and GalC was normalized to actin. Scale bar = 100 μm. * $P < 0.05$, ** $P < 0.001$, significantly different compared to corresponding sham-operated control (S) ($n = 5$/group). *CC*, Corpus callosum; *EC*, external capsule; *IC*, internal capsule; *JPD*, Japanese permissive diet; *Ol*, oligodendrocyte; *SCC*, sham-operated group corpus callosum; *SEC*, sham-operated group external capsule; *SHRSP*, stroke prone spontaneously hypertensive rat; *SIC*, sham-operated group internal capsule; *UCAO*, unilateral carotid artery occlusion.

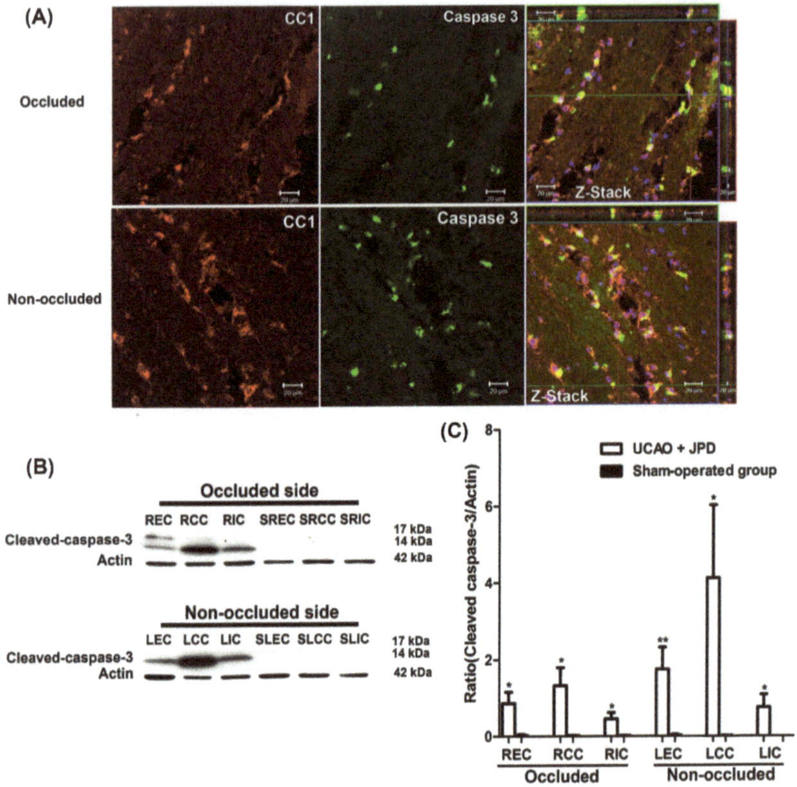

FIGURE 6.3 Ol cell death in SHRSP Apoptotic death of Ols following UCAO/JPD. (A) Cleaved caspase-3 (*green*), and mature Ols (CC1; *red*) positive cells in occluded and nonoccluded sides. *Yellow* stains in Z-stack confocal images indicate co-localization of CC1 and cleaved caspase-3 positive cells. Scale bars = 20 μm. (B) Representative immunoblots of cleaved caspase-3 in occluded (R) and nonoccluded (L) sides in the areas of the EC, CC, and the IC of UCAO/JPD and sham-operated group (S). Actin was used as a loading control. (C) Histogram representing cleaved caspase-3 expression measured using densitometric analysis. Data are expressed as the mean ± S.E.M. and normalized to actin ($n = 6$/group; * $P < 0.05$, ** $P < 0.01$). *CC*, Corpus callosum; *EC*, external capsule; *IC*, internal capsule; *JPD*, Japanese permissive diet; *Ol*, oligodendrocyte; *SHRSP*, stroke prone spontaneously hypertensive rat; *UCAO*, unilateral carotid artery occlusion.

14 W: $P < 0.05$, 15 W: $P < 0.001$), whereas age-matched WKY controls maintained normal pO_2 levels. Additional biochemical studies confirmed the EPR findings and demonstrated a state of hypoxic hypoperfusion in the SHRSP/JPD model. We measured HIF-1α protein expression by Western blot, a transcriptional regulator elevated during hypoxia, and compared it to an actin loading control. Consistent with EPR data, we found HIF-1α up regulation in the dorsal lateral cortex ($P = 0.05$) and hippocampus ($P = 0.046$),

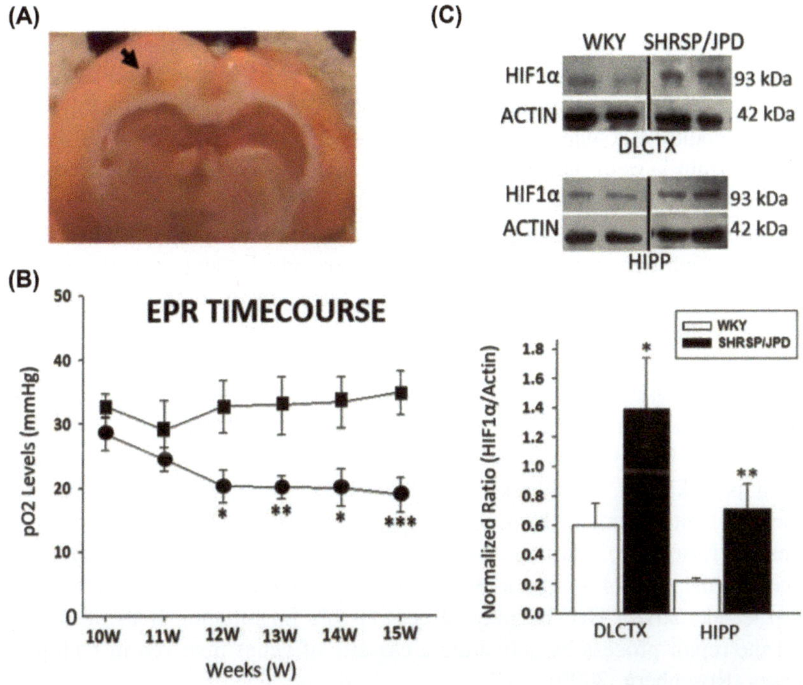

FIGURE 6.4 Cortical and hippocampal hypoxic hypoperfusion in the SHRSP/JPD model. (A). Brain section from SHRSP/JPD confirming the Lithium Phthalocyanine crystal implantation in the right DLCTX (arrow) at 10 W of life. (B). Timecourse of EPR oximetry measurements of interstitial pO_2 readings obtained weekly from 10 to 15 W (WKY, $n = 5$, squares; SHRSP/JPD, $n = 8$, circles) demonstrates a significant drop in pO_2 levels at 12 W (*$P < 0.05$) which remain low until 15 W (**$P < 0.01$, *** $P < 0.001$). (C). Representative WB bands are presented for HIF-1α and normalized to actin loading control. Quantification reveals elevated protein expression for HIF-1α/actin ratio in the DLCTX (*$P = 0.05$; $n = 4$–5/group) and HIPP (**$P = 0.046$; $n = 3$/group) brain regions of SHRSP/JPD compared to controls. WB lanes were run on the same gel but are noncontiguous. *EPR*, Electron paramagnetic resonance; *JPD*, Japanese permissive diet; *SHRSP*, stroke prone spontaneously hypertensive rat; *WB*, Western blot.

where HIF-1α protein expression more than doubled in SHRSP/JPD compared to normotensive WKY controls (Fig. 6.4C; Raz et al., 2019).

Mechanism of white matter injury in SHRSP

It is possible to construct a schematic diagram from the data obtained from SHRSP studies that shows a possible model of the manner in which this complex cascade of MMP expression and activation could proceed during hypoxic injury from initiation of the injury to opening of the BBB mediated by activation of the MMPs. Astrocytes, which are the main source of MMP-2, normally control the integrity of the BBB. With the onset of hypoxia, astrocytes react to the injury signal by activating MMP-2; the membrane bound MMP, MT-MMP, is activated by the formation of the convertase, Furin. MMP-3, which activates MMP-9 is expressed by activated microglia. Both MT-MMP and MMP-3 could be activated by plasmin generated by the action of the plasminogen/plasmin system. Once formed and released into the extracellular space, the MMP-3 attacks most of the components of the extracellular matix, and activates the proMMP-9. Sustained ischemia leads to the amplification of the injury through the production of cytokines, free radicals, and chemokines, and the recruitment of mononuclear cells. The break down of the basal lamina heightens the risk of brain edema and can lead to microbleeds (Rosenberg et al., 2001).

The process of injury is facilitated by the action of the MMPs. However, they are also intimately involved in the repair process; once the injury has occurred the healing process can begin. Because the exact timing of the transition is unknown, the blockade of MMPs can have the unwanted consequence of interfering with recovery. During recovery MMP-9 participates in synapse formation; learning can be blocked when an MMP-9 inhibitor is given. This type of inhibitor would be beneficial during the injury phase (Raz et al., 2018). HIF-1α plays a central role in initiating both the injury and the repair process by activating a cassette of genes involved in both processes (Rosenberg, 2009).

References

Ben-Ari, H., Lifschytz, T., Wolf, G., Rigbi, A., Blumenfeld-Katzir, T., Merzel, T. K., Koroukhov, N., Lotan, A., & Lerer, B. (2019). White matter lesions, cerebral inflammation and cognitive function in a mouse model of cerebral hypoperfusion. *Brain Research, 1711*, 193–201. Available from https://doi.org/10.1016/j.brainres.2019.01.017.

Candelario-Jalil, E., Yang, Y., & Rosenberg, G. A. (2009). Diverse roles of matrix metalloproteinases and tissue inhibitors of metalloproteinases in neuroinflammation and cerebral ischemia. *Neuroscience*, *158*(3), 983–994. Available from https://doi.org/10.1016/j.neuroscience.2008.06.025.

D'Souza, C. A., Mak, B., & Moscarello, M. A. (2002). The up-regulation of stromelysin-1 (MMP-3) in a spontaneously demyelinating transgenic mouse precedes onset of disease. *The Journal of Biological Chemistry*, *277*(16), 13589–13596.

Fredriksson, K., Nordborg, C., Kalimo, H., Olsson, Y., & Johansson, B. B. (1988). Cerebral microangiopathy in stroke-prone spontaneously hypertensive rats – An immunohistochemical and ultrastructural study. *Acta Neuropathologica*, *75*(3), 241–252. Available from https://doi.org/10.1007/BF00690532.

Hainsworth, A. H., Allan, S. M., Boltze, J., Cunningham, C., Farris, C., Head, E., Ihara, M., Isaacs, J. D., Kalaria, R. N., Lesnik Oberstein, S. A. M. J., Moss, M. B., Nitzsche, B., Rosenberg, G. A., Rutten, J. W., Salkovic-Petrisic, M., & Troen, A. M. (2017). Translational models for vascular cognitive impairment: A review including larger species. *BMC Medicine*, *15*(1), 16. Available from https://doi.org/10.1186/s12916-017-0793-9.

Hajdu, M. A., Heistad, D. D., & Baumbach, G. L. (1991). Effects of antihypertensive therapy on mechanics of cerebral arterioles in rats. *Hypertension*, *17*(3), 308–316. Available from https://doi.org/10.1161/01.hyp.17.3.308.

Hattori, Y., Enmi, J.-I., Iguchi, S., Saito, S., Yamamoto, Y., Tsuji, M., Nagatsuka, K., Kalaria, R. N., Iida, H., & Ihara, M. (2016). Gradual carotid artery stenosis in mice closely replicates hypoperfusive vascular dementia in humans. *Journal of the American Heart Association*, *5* (2). Available from https://doi.org/10.1161/JAHA.115.002757.

Ihara, M., Tomimoto, H., Kinoshita, M., Oh, J., Noda, M., Wakita, H., Akiguchi, I., & Shibasaki, H. (2001). Chronic cerebral hypoperfusion induces MMP-2 but not MMP-9 expression in the microglia and vascular endothelium of white matter. *Journal of Cerebral Blood Flow and Metabolism*, *21*(7), 828–834. Available from https://doi.org/10.1097/00004647-200107000-00008.

Jalal, F. Y., Yang, Y., Thompson, J., Lopez, A. C., & Rosenberg, G. A. (2012). Myelin loss associated with neuroinflammation in hypertensive rats. *Stroke: A Journal of Cerebral Circulation*, *43* (4), 1115–1122. Available from https://doi.org/10.1161/STROKEAHA.111.643080.

Kitamura, A., Fujita, Y., Oishi, N., Kalaria, R. N., Washida, K., Maki, T., Okamoto, Y., Hase, Y., Yamada, M., Takahashi, J., Ito, H., Tomimoto, H., Fukuyama, H., Takahashi, R., & Ihara, M. (2012). Selective white matter abnormalities in a novel rat model of vascular dementia. *Neurobiology of Aging*, *33*(5). Available from https://doi.org/10.1016/j.neurobiolaging.2011.10.033, 1012.e25–35.

Nakaji, K., Ihara, M., Takahashi, C., Itohara, S., Noda, M., Takahashi, R., & Tomimoto, H. (2006). Matrix metalloproteinase-2 plays a critical role in the pathogenesis of white matter lesions after chronic cerebral hypoperfusion in rodents. *Stroke: A Journal of Cerebral Circulation*, *37*(11), 2816–2823. Available from https://doi.org/10.1161/01.STR.0000244808.17972.55.

Okamoto, K., & Aoki, K. (1963). Development of a strain of spontaneously hypertensive rats. *Japanese Circulation Journal*, *27*, 282–293.

Raz, L., Yang, Y., Thompson, J., Hobson, S., Pesko, J., Mobashery, S., Chang, M., & Rosenberg, G. (2018). MMP-9 inhibitors impair learning in spontaneously hypertensive rats. *PLoS One*, *13*(12), e0208357. Available from https://doi.org/10.1371/journal.pone.0208357.

Raz, L., Bhaskar, K., Weaver, J., Marini, S., Zhang, Q., Thompson, J. F., Espinoza, C., Iqbal, S., Maphis, N. M., Weston, L., Sillerud, L. O., Caprihan, A., Pesko, J. C., Erhardt, E. B., & Rosenberg, G. A. (2019). Hypoxia promotes tau hyperphosphorylation with associated neuropathology in vascular dysfunction. *Neurobiology of Disease*, *126*, 124–136. Available from https://doi.org/10.1016/j.nbd.2018.07.009.

Rosenberg, G. A., Cunningham, L. A., Wallace, J., Alexander, S., Estrada, E. Y., Grossetete, M., Razhagi, A., Miller, K., & Gearing, A. (2001). Immunohistochemistry of matrix metalloproteinases in reperfusion injury to rat brain: Activation of MMP-9 linked to stromelysin-1 and microglia in cell cultures. *Brain Research*, *893*(1–2), 104–112.

Rosenberg, G. A. (2009). Matrix metalloproteinases and their multiple roles in neurodegenerative diseases. *The Lancet Neurology*, *8*(2), 205–216. Available from https://doi.org/10.1016/S1474-4422(09)70016-X.

Shibata, M., Ohtani, R., Ihara, M., & Tomimoto, H. (2004). White matter lesions and glial activation in a novel mouse model of chronic cerebral hypoperfusion. *Stroke: A Journal of Cerebral Circulation*, *35*(11), 2598–2603. Available from https://doi.org/10.1161/01.STR.0000143725.19053.60.

Sood, R., Yang, Y., Taheri, S., Candelario-Jalil, E., Estrada, E. Y., Walker, E. J., Thompson, J., & Rosenberg, G. A. (2009). Increased apparent diffusion coefficients on MRI linked with matrix metalloproteinases and edema in white matter after bilateral carotid artery occlusion in rats. *Journal of Cerebral Blood Flow and Metabolism*, *29*(2), 308–316. Available from https://doi.org/10.1038/jcbfm.2008.121.

Wakita, H., Tomimoto, H., Akiguchi, I., & Kimura, J. (1994). Glial activation and white matter changes in the rat brain induced by chronic cerebral hypoperfusion: an immunohistochemical study. *Acta Neuropathologica*, *87*(5), 484–492. Available from https://doi.org/10.1007/BF00294175.

Wakita, H., Tomimoto, H., Akiguchi, I., Matsuo, A., Lin, J. X., Ihara, M., & McGeer, P. L. (2002). Axonal damage and demyelination in the white matter after chronic cerebral hypoperfusion in the rat. *Brain Research*, *924*(1), 63–70.

Yamori, Y., Horie, R., Handa, H., Sato, M., & Fukase, M. (1976). Pathogenetic similarity of strokes in stroke-prone spontaneously hypertensive rats and humans. *Stroke: A Journal of Cerebral Circulation*, *7*(1), 46–53. Available from https://doi.org/10.1161/01.STR.7.1.46.

Chapter 7

Molecular pathways

Introduction

The growing number of people worldwide over the age of 65 has created challenges for governments who need to find ways to prepare for the large increase expected in people with dementia, which will overwhelm societies' ability to pay for the anticipated increased costs of caring for them. The increase in longevity has been called a "Silver Tsunami," and it is the main cause for the dramatic increase in the number of people with dementia. The percentage of people with dementia doubles every 5 years, and the improved care for heart disease and cancer, the two major causes of death, along with the lack of a disease modifying treatment for dementia, makes the brain the most vulnerable organ facing the consequences of aging. In 1950 5% of the world's population was over 65 years old. This is projected to increase to 16% by the year 2050. At the same time, those between 20 and 64 years of age will reduce from 44% to 27%, decreasing the number of people able to support the elderly. The funding for aging and dementia research in national health institutes has greatly increased in an attempt to find a way to slow the inevitable progress of aging diseases. Worldwide, life expectancy continues to rise. By 2020, for the first time in history, the number of people aged 60 years and older will outnumber children younger than 5 years. By 2050, the world's population aged 60 years and older is expected to total 2 billion, up from 841 million today. Eighty percent of these older people will be living in low- and middle-income countries. The increase in longevity, especially in high-income countries, has been largely due to the decline in deaths from cardiovascular disease, mainly because of simple, cost-effective strategies to reduce tobacco use and high blood pressure, and improved access and effectiveness of health interventions. However, although people are living longer, they are not necessarily healthier than before—nearly a quarter (23%) of the overall global burden of death and illness is in people aged over 60, and much of this burden is attributable to long-term illness caused by diseases such as cancer, chronic respiratory diseases, heart disease, musculoskeletal diseases (such as arthritis and osteoporosis), and mental and neurological disorders. This long-term burden of illness and diminished well-being affects patients, their families, health systems, and economies, and is forecast to accelerate. For example, the latest estimates indicate that the number of people with dementia is expected to rise from 44 million now, to 135 million by 2050.

Neuroinflammation in Vascular Dementia. DOI: https://doi.org/10.1016/B978-0-12-823455-6.00002-X

As the support of the amyloid hypothesis for Alzheimer's disease (AD) wanes, there is an increase in a number of hypotheses that had been overlooked in the enthusiasm for amyloid as a cause of AD. The impact of aging has created new interest in the science of aging. Although it is not possible to overcome the consequences of aging, but there is a growing realization that many of the disease processes of aging can be ameliorated. Vascular disease is a major contributor to cognitive decline with aging, and it begins to exert negative effects in midlife. This makes it possible with public health efforts to reduce the burden of vascular disease by changing lifestyles. As more people improve their diets, increase exercise, and treat hypertension and diabetes, the number of people that will develop dementia, which has already begun to decline, will be positively impacted. Aging of the vascular system is inevitable with increased age. The loss of elasticity and the thickening of the vessel and surrounding matrix results in a stiffer vessel with less ability to make changes in caliber with changes in position. Aging blood vessels contribute to the reduced cerebral blood flow (CBF).

Burden of disease in normals

Accumulation of abnormal proteins is one of the inevitable consequences of aging. These misfolded proteins can be tolerated and removed by normal waste disposal processes. However, as the removal mechanisms are overwhelmed, the abnormal proteins activate detrimental molecular pathways, particularly those involved in inflammation. A number of studies have assessed the burden of these proteins in normal individuals; these studies are important in understanding the role of misfolded proteins in patients with dementia. In large series of autopsy cases of patients with dementia, there are numbers of cognitively normal (CN) individuals. Four such studies have been combined to create a large database of normal individuals that had normal cognition within 2 years of the autopsies. The complete set included 424 brains out of 1672 autopsies obtained from four series: the Adult Changes in Thought study, the Honolulu-Asia Aging Study, the Nun Study, and the Oregon Brain Aging Study (Sonnen et al., 2011). They found that 47% of CN cases had moderate or frequent neuritic plaque (NP) density, and of these 6% also had Braak stage V or VI for neurofibrillary tangles (NFTs). Furthermore, 15% of CN cases had medullary Lewy Body Disease (LBD); 8% had nigral and 4% isocortical LBD. The presence of any cerebral microinfarcts (CMIs) was identified in 33% and high-level CMIs in 10% of CN individuals. The investigators concluded that (1) AD, microvascular brain injury (μVBI), and LBD were prevalent among CN older adults; and (2) the burden of these three diseases varied widely, and co-morbidity among these three diseases is quite variable among individuals. This study was unusual in that others have reported on neuropathologic evidence of one or two of these diseases, but few had such a comprehensive study, involving multiple abnormal proteins. In this group of CN with a wide age range, the number of pathological processes that they found increased with age. A significant number of CN had one or more pathological protein in their brain at the time of death. The accumulation of abnormal proteins over time confounds attempts to attribute an individual patient with a single pathological entity. This is particularly true for AD and vascular disease, which are commonly found together.

Consequences of abnormal proteins in the brain

As the abnormal misfolded proteins increase in the brain under pathological conditions, they create an inflammatory environment. The amyloid-related proteins have been shown to have deleterious effects by activating microglia and inducing a series of molecules that increase inflammation. Aggregated, amyloidβ (Aβ) is sensed by microglial pattern-recognition receptors, producing a pathological innate immune activation with subsequent production of inflammatory mediators. Activation of the NACHT-, LRR-, and pyrin (PYD)-domain-containing protein 3 (NLRP3) inflammasome, a central sensor for danger signals, has recently been documented in the brains of patients with AD and in *APPSwe/PSEN1dE9* (also known as APP/PS1) transgenic mice. Genetic deficiency in NLRP3 or caspase-1 both protect aged APP/PS1 mice from microglial IL-1β production, Aβ-related pathology and development of cognitive decline. Once aggregated, Aβ is sensed by microglial pattern-recognition receptors leading to pathological innate immune activation and subsequent production of inflammatory mediators. After activation, NLRP3 recruits the adapter protein apoptosis-associated speck-like protein containing a CARD (caspase recruiting domain) (ASC) via interactions with the PYD domain of ASC, which triggers ASC helical fibrillar assembly. ASC fibrils then recruit the effector caspase-1 via CARD interactions leading to autoproteolytic activation and subsequent assembly of ASC fibrils into large paranuclear ASC speck. In addition to causing proinflammatory IL-1β cytokine activation and release, NLRP3, inflammasome activity also results in the release of assembled ASC specks (Venegas et al., 2017).

In the brains of Parkinson's disease patients, there is an accumulation of α-synuclein, which is another misfolded protein that increases with age and similarly to amyloid results in inflammation. The brain is vulnerable to oxidative stress in comparison to peripheral organs due to high levels of polyunsaturated fats and relatively low antioxidant activity. There is a bi-directional relationship between α-synuclein oligomerization and the generation of reactive oxygen species (ROS). Microglial cells contribute to oxidative stress following α-synuclein accumulation. When microglial cells engulf α-synuclein in an attempt to clear it from the extracellular space, either as a result of apoptotic neuron death, or from mechanisms of cellular release, α-synuclein-induced activation of microglia leads to multiple proinflammatory changes that ultimately are neurotoxic, including increased production of ROS (Rocha et al., 2018).

A fourth abnormal protein, TDP-43, has been described in patients with hippocampal sclerosis. It is found in brains of elderly patients associated with arteriolosclerosis, and is specific to hippocampal sclerosis in aging because brains with AD pathology did not show the same morphological alterations. There are changes in brain arterioles that impact multiple brain areas and contribute to hippocampal sclerosis (Neltner et al., 2014). The blood supply to the hippocampus is a watershed pattern, making it vulnerable to reduced CBF. Arteriolosclerosis limits the blood flow to the watershed areas (Blevins et al., 2021).

The cumulative effect of these abnormal proteins probably acts synergistically to accelerate the activation of microglia and the recruitment of macrophages. Tests to detect these proteins are available only for amyloid, tau, and α-synuclein. Misfolded proteins stimulate inflammation, which is further accelerated by vascular disease. When the number of pathological proteins increases, the cumulative effect is to increase inflammation. Inflammation attacks cell membranes, while in blood vessels it increases vascular permeability, resulting in the leakage of substances from the systemic circulation that are toxic to brain cells. Blood–brain barrier (BBB) disruption is one of the main pathological mechanisms in patients with dementia.

Neurovascular unit

A number of cells act collectively to separate blood and brain tissue. At each site of contact between the blood and brain cells, proteins self-assemble to form tight junctions. These tight junction proteins are found in the cerebral endothelial cells, where they are most abundant, in the apical cells of the choroid plexus, and in the arachnoid granulations. Around the blood vessels several cells act together to form a neurovascular unit (NVU): basal lamina forms the layer next to the endothelial cells, acting as a charge and molecular size filter. Embedded in the basal lamina are pericytes that have properties of both macrophages and smooth muscle cells. The next layer, outside the basal lamina, are astrocytic foot processes. The astrocytes have aquaporin molecules that are water channels for edema to enter and leave the brain. Because of the extensive network of blood vessels in the brain, the endothelial tight junction proteins are the major site of the BBB. The molecules of the tight junctions, JAM (junctional adhesion molecules), occludin, claudins, and zona occludins, act to "zip lock," forming the first line of defense to prevent molecules in the blood from entering the brain (Hawkins & Davis, 2005).

As more is learned about the evolution of the NVU, a more complex picture is emerging that shows the evolution of the endothelial cells and the glial cells in different species. Molluscs have multiple layers of cells without specific tight junctions. Insects have a glial barrier, but lack one at the level of the endothelial cells. The family of elasmobranches that include the sharks has developed endothelial cells, but the tight junctions remain in the glial cells. Humans have lost the glial barrier and developed the barrier at the endothelial cell. The NVU in the mammalian brain has evolved over millions of years to reach the form found in humans (Fig. 7.1). Homo sapiens have an endothelial cell barrier with tight junction proteins. The tight junctions were absent in the molluscs and present in the glial cells in the insects and the elasmobranches, appearing in the endothelial cells in the mammals (Abbott et al., 1985).

Evolution did not follow a straight line from the glial barrier to the endothelial barrier with branches at several species. In one scheme, the barrier located at the glial level in invertebrates continues into the ancestral vertebrate(s), then shifts to an endothelial location in the course of evolution, as a result of significant selective advantage. Advantages could include the division of labor between endothelium and glia. In mammals, the endothelium became specialized for regulating permeability and transport at the blood–brain interface, while glial cells became more involved with neuronal support and induction/communication within the gliovascular unit (Bundgaard & Abbott, 2008).

The BBB is an essential regulator of blood–brain exchange of nutrients and waste removal that operates on several distinct levels. There are a number of ways for substances to cross the BBB besides through an open tight junction. Carrier-mediated transport mechanisms facilitate glucose and amino acid entry into brain. Small and/or lipophilic molecules can bypass the tight

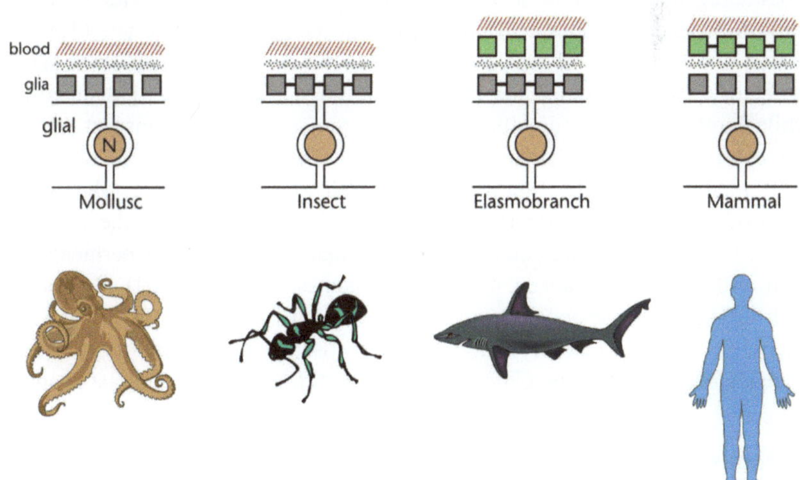

FIGURE 7.1 **Evolution of the (BBB.** Molluscs lack a BBB and the blood is contiguous with the glia cells that lack tight junctions. Insects have the tight junctions in the glial cells. In the elasmobranchs, a layer of endothelial cells are present with the tight junction remaining in the glial cells. Mammals have shifted the tight junctions to the endothelial cells and the glial cells no longer have tight junctions. *BBB*, blood-brain barrier. *(Figure derived from Abbott et al 1985).*

junctions but are rapidly degraded by intracellular enzymes and actively pumped back into the bloodstream by P-glycoprotein 1 (permeability glycoprotein, Pgp) also known as multidrug resistance protein 1 (MDR1) is an important protein of the cell membrane that pumps many foreign substances out of cells; it is an ATP-dependent efflux pump with broad substrate specificity (Choi & Yu, 2014). The clefts between endothelial cells prevent large molecules from crossing even when disrupted. However, they can use pinocytosis, which is movement through the endothelial cell membranes to move from blood to brain. During experimental hypertensive encephalopathy, cerebral cortical arterioles in focal neocortical areas develop increased permeability to plasma proteins and protein tracers. The mechanism underlying this increased permeability is pinocytosis. During acute hypertension in rats, cerebral cortical arterioles develop increased permeability by pinocytosis, involving transendothelial vesicular movement (Nag et al., 1979).

Although the BBB is a highly effective gatekeeper, substances that are either lipophilic or smaller than 400 Da may effectively bypass the physical barrier. The metabolic barrier eliminates such chemicals that would otherwise move from the blood into the central nervous system. As the primary route of drug metabolism, p450-related cytochrome enzyme oxidizes unwanted substances within the cytoplasm of endothelial cells. Enzymatic activity within the BBB facilitates the transport of essential substances from blood to brain. Through its transpeptidase activity, gamma-glutamyl transpeptidase (GGT) assists in the transfer of amino acids across the BBB. By modulating pyroglutamate levels, GGT indirectly stimulates a wide array of sodium-dependent amino acid transporters at the plasma membrane (Tam & Watts, 2010). Routes of transport appear to be different for different sized molecules. Smaller molecules can cross through a disrupted tight junction, while large ones, such as albumin require transendothelial transport by the pinocytotic route. This differential transport dependent on molecular size most likely explains the finding of increased albumin in the cerebrospinal fluid (CSF) in some patients with a relatively intact BBB to the contrast agent, Gadolinium. This could explain the marked increase in permeability seen with contrast-enhanced magnetic resonance imaging (MRI) in patients with inflammatory lesions due to multiple sclerosis with low levels of CSF albumin. While in elderly patients with vascular dementia, there is an increase in albumin in the CSF with only a slight leak of the contrast agent.

Methods to measure permeability at the neurovascular unit

A wide variety of methods both quantitative and qualitative are used to measure the permeability of the BBB. Some are only available for animal studies since they require removal of the brain; in animals it is possible to use dyes, such as trypan blue, and isotopes. Radiolabeled sucrose is a convenient tracer to use since it is mainly excluded from entry to the brain; it can show low levels of leakage. The tracer is injected intravenously and uptake into the brain measured after 20–30 minutes by removing the brain and measuring the amount of the tracer in the brain and blood (Ohno et al., 1978). While alcohols rapidly cross the BBB, water crosses more slowly. In a study in rhesus monkeys, using labeled water ($H_2{}^{15}O$), the labeled water did not equilibrate freely when cerebral blood flow (CBF) exceeded 20 (mL/100 g)/min; at normal CBF [50 (mL/100 g)/min] only 90% of the injected bolus freely equilibrated with brain, and this value progressively declined as CBF increased (Raichle et al., 1974). In human studies, it is routine to use iodine-based tracers, which are dense to x-rays as the contrast agents for computed tomography (CT), while gadolinium-based tracers that change characteristics of water molecules are used in MRI.

Molecular expression patterns in endothelial and mural cells vary at different levels of the vascular tree creating heterogeneity (zonation) at different levels. Arteries penetrate into the brain from surface bringing with them a layer of pial cells for a short distance. This creates the Virchow–Robin space at the level of penetrating arteries, where endothelial cells form the inner layer of the vessel wall. The Virchow–Robin space is between the pia and the glia limitans formed by astrocytic endfeet. At the arteriolar level, smooth muscle cells form a single layer, whereas the endothelial layer is continuous with the endothelium of penetrating arteries and capillaries. Deeper in the brain as the vessels move away from the surface, the capillaries, pericytes, and endothelial cells share a basal lamina. Both the arteriolar and capillary vessel wall is covered by astrocytic endfeet. Smooth muscle cells, pericytes, and astrocytes all have neuronal innervation. The BBB is a series of layers with the innermost layer formed by endothelial cells connected together by tight junction proteins, forming a monolayer of tightly sealed endothelial cells extending along the vascular tree and expressing low paracellular and transcellular permeability at the level of brain capillaries (Sweeney et al., 2019).

Measurements of transport at the neurovascular unit

Permeability of the BBB can be quantitatively measured with gadolinium-based contrast agents. The basic method was originally used with radiolabeled tracers and autoradiograms; following a given amount of time after injection of the tracer, the brain is removed, sectioned, and the tracer concentration in the brain was measured in autoradiographs. Knowing the concentration in the blood, a transfer constant can be calculated by the use of the Patlak graphical method. The autoradiographic method was initially transferred to animals using MRI to substitute for autoradiograms (Ewing et al., 2003). The transfer to human studies used the same methodology as in those in animals (Taheri et al., 2011). This method is called dynamic contrast-enhanced MRI (DCE-MRI). Other methods are also used, but the DCE-MRI is the main one (Raja et al., 2018), and it has been recommended for human studies in a consensus report on optimal methods to quantify the permeability of the blood vessels (Smith et al., 2019).

Two main methods have been used to measure BBB permeability in humans: the ratio of albumin in the CSF to that in the blood, which is called the albumin index (Alb_{ind}) and the MRI method, DCE-MRI. Initial interest in measurement of BBB permeability in humans came from studies in patients with multiple sclerosis. In order to quantify the presence of abnormal proteins in the CSF, the amount of the protein derived from the blood needed to be known. Since very little albumin crosses normally into the CSF, the presence of albumin in the CSF was considered an indication of the leakiness of the BBB, and it was used to index CSF proteins, such as IgG, in order to determine the amount produced by the brain (Tourtellotte et al., 1985). Because lumbar puncture is invasive, and carries a small risk of complications, the measurement of the Alb_{ind} waned, and in many research studies has been replaced with DCE-MRI. The measurement of permeability with the albumin index and the DCE-MRI show different results. Albumin index is a global measure that indicated movement of albumin into the brain, but has no regional information, while DCE-MRI provides a pixel-by-pixel map of the sites of increased permeability. The albumin is a very large molecule, while gadolinium is a small one. Measurements of permeability in the same patient show that the two measures are not equivalent. DCE-MRI shows marked increased in permeability during an inflammatory episode in multiple sclerosis, suggesting the inflammation has opened the tight junctions. In vascular cognitive impairment, where the DCE-MRI shows very low permeability and the albumin index is elevated, it is likely that pinocytosis is the main route for albumin entry into the brain.

Disruption of the neurovascular unit in dementia

Alzheimer's disease

The first measurements of BBB in dementia were done with the albumin index (Alb_{ind}); patients with AD and vascular cognitive impairment were studied and the vascular disease patients had consistently elevated Alb_{ind} as opposed to those with AD, which had inconsistent patterns (Wallin et al., 1990). These early studies separated patients clinically, and from subsequent studies, it is highly likely that some of these patients had dual pathologies with both Alzheimer's proteins and vascular changes in the brain, which may explain the variable results. Albumin is a large molecule suggesting that the transport across the blood vessels may have involved pinocytosis rather than a disruption of the tight junctions with the albumin moving through the blood vessels.

Neurodegeneration has a predilection for certain regions. The reason for the selective vulnerability is not known. Among the most vulnerable neurons in AD are the large pyramidal neurons in the cortex, particularly association neurons with long projections. Examples of such neurons are those in the entorhinal cortex layer II which provides the major cortical input to the hippocampus through the perforant pathway, and those in the subiculum which represent the major output of the hippocampus to the prefrontal cortex and other brain regions. The underlying factors for the vulnerability of these large-sized neurons are thought to be their high demand for energy that leads to high levels of oxidative phosphorylation. Oxidative phosphorylation depends on the activity of the electron transport chain enzymes, and hyperactivation of these enzymes is known to lead to oxidative stress, which could modify proteins, DNA, and lipids, resulting in neuronal injury. In addition, the relatively large surface area of these neurons may increase exposure to extracellular toxic agents while their long processes would put increased demands on axoplasmic transport over long distances. Finally, neurons with long processes also have high levels of neurofilaments whose protein subunits tend to form aggregates during cellular stress (Wang et al., 2010). AD affects parts of the temporal lobe, such as the hippocampus, the cingulate, and the dorsal anterior cingulate cortex. The hippocampal region is where β-amyloid deposition begins, and when pTau builds up in neurons, NFTs are formed, leading to cell death. This leads to inflammation and disruption of the BBB related to expression of matrix metalloproteinases (MMPs). High-resolution MRI showed regional permeability differences between CA areas in the hippocampus of patients with mild cognitive impairment (MCI) and AD compared to normal controls. Increased permeability was seen in both CA1 and CA3; white matter did not show an increase in permeability (Fig. 7.2) (Montagne et al., 2015). Others have observed increased leakiness in the gray matter in AD, but have not separated the regions of the gray matter involved in as great a detail (van de Haar et al., 2016).

The *APOE* gene has been implicated in vulnerability of the BBB. Human apolipoprotein E has three isoforms: *APOE2, APOE3*, and *APOE4*; *APOE4*is a major genetic risk factor for late-onset AD. Neurovascular dysfunction is present in normal *APOE4* carriers: mice without the *APOE* gene show BBB breakdown. Expression of *APOE4*, but not *APOE2* and *APOE3*, leads to BBB breakdown by activating a proinflammatory cyclophilin A (CyPA)–nuclear factor-κB–matrix-metalloproteinase-9 pathway in pericytes (Bell et al., 2012). CyPA is a ubiquitously distributed protein belonging to the immunophilin family that regulates protein folding and trafficking as an intracellular protein. Current research in animal models and humans has provided compelling evidence supporting the critical function of CyPA in several human diseases (Fig. 7.3) (Nigro et al., 2013).

Vascular cognitive impairment

While gray matter has a relatively uniform pattern of injury in highly selective vulnerable areas, such as the hippocampus, the diverse disease processes

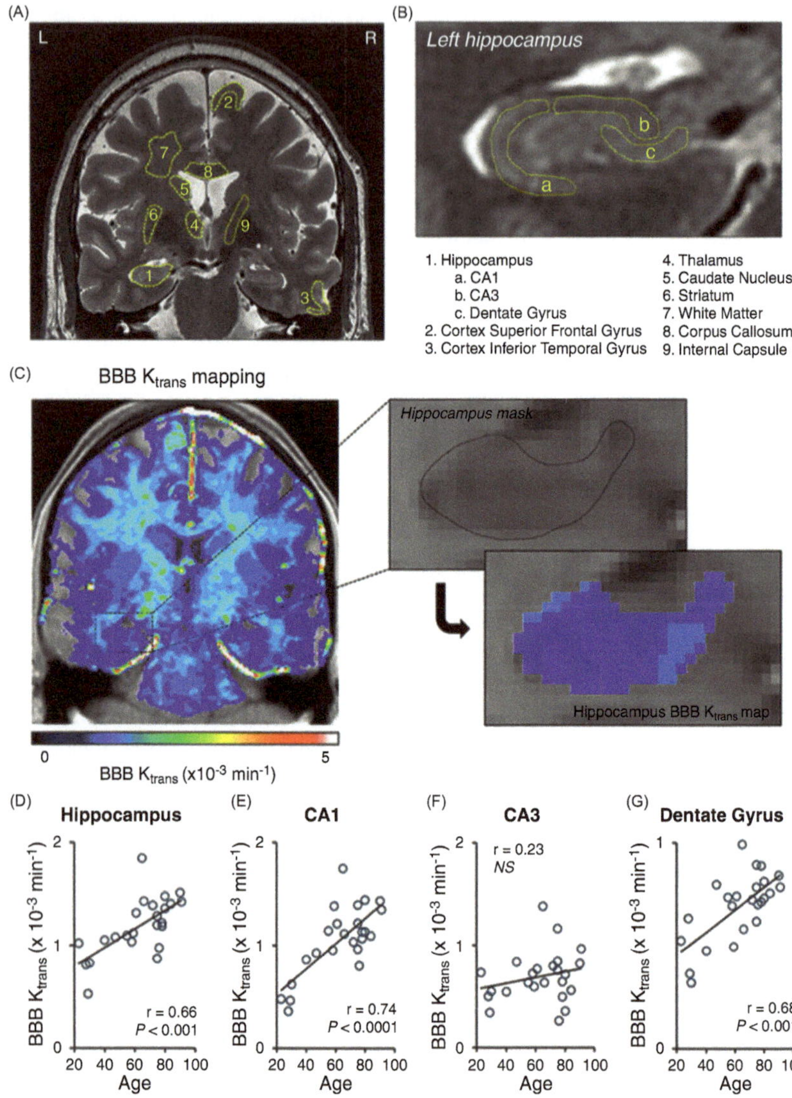

FIGURE 7.2 (A–B) A dynamic contrast enhanced magnetic resonance imaging was used to quantify the BBB regional permeability K_{trans} constant in 12 regions-of-interest in the gray and white matter in the living human brain (A) including subdivisions of the hippocampus (B). (C) Representative BBB K_{trans} maps of the whole brain (left) and the hippocampus (right) in a young participant with no cognitive impairment. (D–G) Age-dependent increase in the BBB permeability K_{trans} constant in the entire hippocampus, its CA1 region and dentate gyrus, but not CA3 region. Single data points for the K_{trans} constant from 24 individuals with no cognitive impairment (both genders, ages 23–91) were plotted against age; *r* = Pearson's coefficient; °*P*, significance; *NS*, nonsignificant. *BB*, Blood–brain barrier.

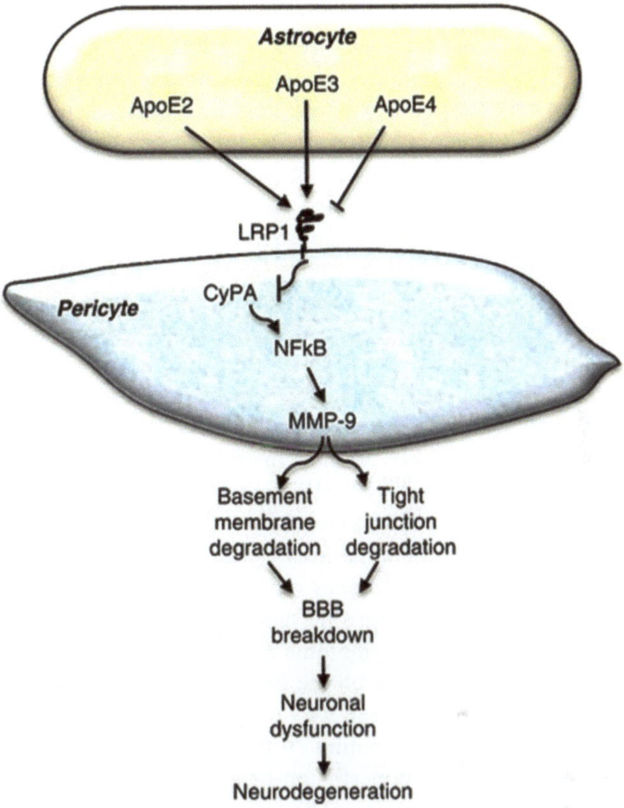

FIGURE 7.3 **Cyclophilin A mediates BBB opening.** Cyclophilin A promotes neurodegeneration. ApoE2 and ApoE3 secreted by astrocytes, but not ApoE4, stimulates the low-density LRP1 on pericytes. This event suppresses the proinflammatory pathway orchestrated by Cyclophilin A that leads to NF-kB and MMP-9 activation. As a consequence, the MMP-9-mediated degradation of tight junction and basement membrane proteins that causes BBB breakdown and neuronal dysfunction is inhibited. *BBB*, Blood–brain barrier; *LRP1*, lipoprotein receptor-related protein 1. *From Nigro, P., Pompilio, G., & Capogrossi, M. C. (2013). Cyclophilin A: A key player for human disease.* Cell Death & Disease, 4, *e888. https://doi.org/10.1038/cddis.2013.410.*

that affect the white matter defy lumping all injury pathways together. The lack of specific areas of degeneration in vascular dementia makes it difficult to use the structural MRI to separate vascular disease from other forms of dementia, increasing the complexity of following regional changes over time. As opposed to AD, the cognitive loss in vascular dementia does not follow specific patterns, but affects different cognitive functions individually. This creates problems when treatment trials are contemplated. To assess a treatment in a clinical trial, it is necessary to understand the natural history of the disease so that any improvement due to the treatment can be evaluated. Furthermore, in a clinical trial in multi-infarct dementia and post-stroke dementia, both of which progress sporadically, a larger number of patients are required than needed in a disease process with a relatively well-known pattern of progression. In vascular cognitive impairment due to multiple strokes, there is the lack of a specific pattern of disease progression, due to the sporadic nature of strokes; a much larger number of patients are needed than in a relatively progressive disease process. One study showed that the changes in brain volume as measured with T2 white matter hyperintensity volume, lacunes, and white matter damage on diffusion tensor imaging (DTI) were detectable over 3 years in all of these MRI markers. Using these markers could significantly reduce the size of trials to screen treatments for efficacy in small vessel disease (Benjamin et al., 2016). This suggests that small vessel disease should be considered separately from large vessel disease (Fig. 7.4).

Small vessel disease causes lacunar infarcts and Binswanger's disease (BD), which can be separated with MRI and eventually at autopsy. In BD, the primary injury seen on MRI involves the white matter, while in lacunar infarcts it is also seen in some gray matter regions, such as the basal ganglia. While both show differences on MRI and pathologically, the BD form primarily involves damaging the white matter, sparing gray matter. Clinically, BD and AD have symptoms that overlap, making the separation of AD and vascular disease on the basis of clinical information difficult. The vulnerability of the deep white regions has multiple causes; the main one being the end artery nature of the cerebrovasculature, exposing it to repeated episodes of hypoperfusion or more rarely to a major loss of CBF during serious heart disease. The cascade of molecular events that follow hypoxic hypoperfusion are particularly likely to affect the blood vessels in the white matter. Although differences in the incidence of multiple lacunar strokes versus more diffuse hypoxia suggest that small strokes are less likely to lead to extensive white matter damage than more diffuse processes.

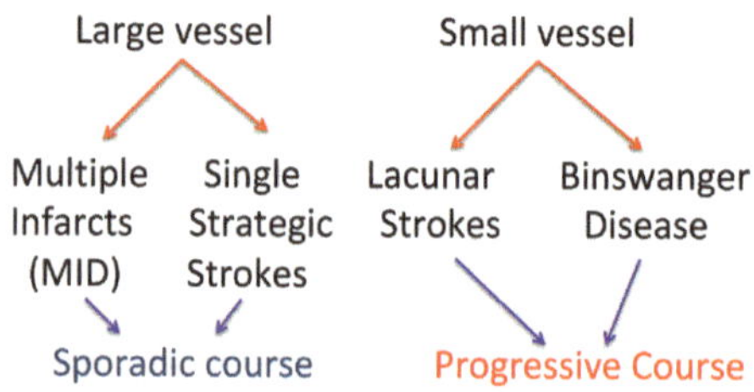

FIGURE 7.4 **Classification of patients into large and small vessel disease.** Large vessel disease can be divided into MID and single strategic strokes; these individual strokes occur sporadically. Small vessel disease presents as lacunar strokes or Binswanger's disease and follows a more progressive course.*MID*, multi-infarct dementia.

Structural differences between the gray and white matter offer further insight into the patterns of injury. Gray matter has a dense neuropil that severely restricts molecular movement between cells. The size of the extracellular space in gray matter is estimated to be 15%; that space is filled with glycosaminoglycans primarily hyaluronic acid. These molecules have large water domains that maintain the size of the space, but restrict movement of other molecules (Srivastava et al., 2020). As opposed to gray matter, the fiber tracts that form the white matter act as conduits for fluid movement; production of interstitial fluid from the secretory action of brain capillaries moves fluid from the deep white matter into the CSF for absorption back into the blood (Rosenberg et al., 1980). Transport of fluid also occurs along perivascular pathway; both the bulk flow between fiber tracts and the perivascular pathway act as the lymphatic system for the brain (Rasmussen et al., 2018).

Binswanger in his original report emphasized the central role of the arterioles that penetrate into the deep white matter (Román, 1987). When the arterioles are injured, blood products such as immunoglobulins and fibrinogen leak into the tissue. Albumin, which is large and normally contained in the blood, also leaks into the CSF (Akiguchi et al., 1998). MRI has added new insights into the role of the BBB in white matter because of its ability to visualize brain with high spatial resolution. Because of the low vascular permeability and low blood flow in the white matter, early attempts at detecting opening of the BBB with MRI were unsuccessful. The breakthrough that allowed accurate measurements of BBB came from autoradiographic studies with radiolabeled tracers that crossed the blood vessels. The initial measurements done with autoradiograms used the Patlak Graphical Method (Patlak & Blasberg, 1985). The Patlak Graphical Method provided quantitative permeability measurements, but the need to remove the brain for the autoradiogram measurements limited it to animal studies (Ewing et al., 2003). When the method was translated to small animal MRI with the use of the contrast agent, gadolinium, further modifications for use in human studies became possible with a standard 1.5 or 3 T MRI (Taheri et al., 2011). This animal method was adopted to MRI in animal studies that proved its reliability (Knight et al., 2009). In humans the contrast agent is infused intravenously for 20 min with frequent sampling by special T1-weighted MRI sequences. The time is too short to allow for significant backflow of contrast, which simplified the transport equations to one direction. The concentration of the contrast in the blood, which is needed to calculate a transfer constant, can be obtained from the blood in the sagittal sinus. Dynamic contrast enhanced-magnetic resonance imaging (DCE-MRI) is the optimal MRI method to quantitate BBB permeability, particularly when there is very low levels of transport (Raja et al., 2018). DCE-MRI has become the preferred method of BBB permeability measurement for research studies (Smith et al., 2019).

DCE-MRI in cerebral small vessel disease

Although blood vessels in the white matter have low CBF and low permeability, there is a large homogeneous amount of tissue to sample, which was the reason that early studies of BBB permeability in dementia focused on the white matter. In addition, the gray matter of the cortex has deep CSF containing sulci next to the gray matter, which confounds the measurements of BBB permeability in the gray matter. The hippocampus provides a better site for measurements of permeability in the gray matter because of its larger size and greater amount of gray matter.

Patients with white matter lesions have increased permeability compared with controls (Taheri et al., 2011; Wong et al., 2019). The regions with the increased permeability are the white matter hyperintensities on fluid attenuated inversion recovery (FLAIR) MRI (Fig. 7.5). DCE-MRI has the advantage of showing the regional distribution of the increased permeability; the major sites of increased permeability in the white matter are around the white matter hyperintensities (the penumbra) and in the normal-appearing white matter

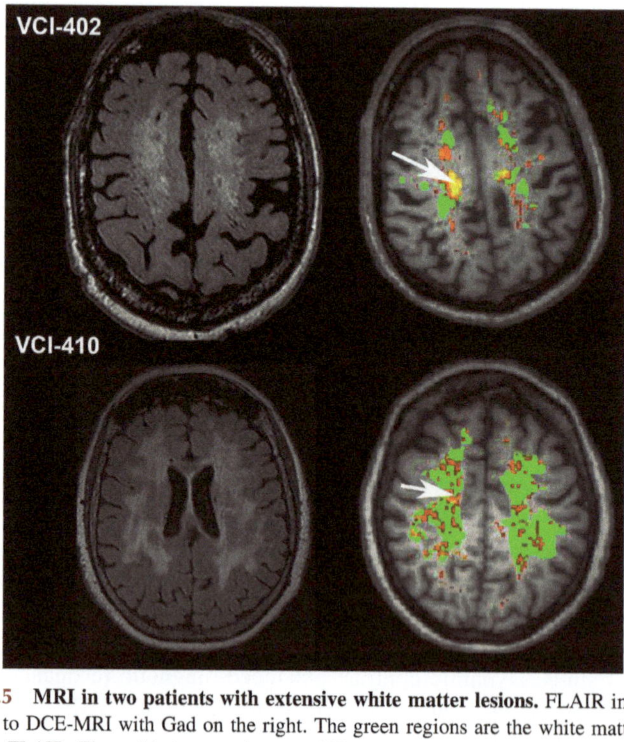

FIGURE 7.5 **MRI in two patients with extensive white matter lesions.** FLAIR image on the left related to DCE-MRI with Gad on the right. The green regions are the white matter hyperintensities on FLAIR. The red is increased permeability, and the red if the most increased permeability. The arrows highlight areas with high permeability (yellow). *DCE-MRI*, dynamic contrast enhanced MRI; *FLAIR*, Fluid attenuated inversion recovery; *MRI*, Magnetic resonance imaging.

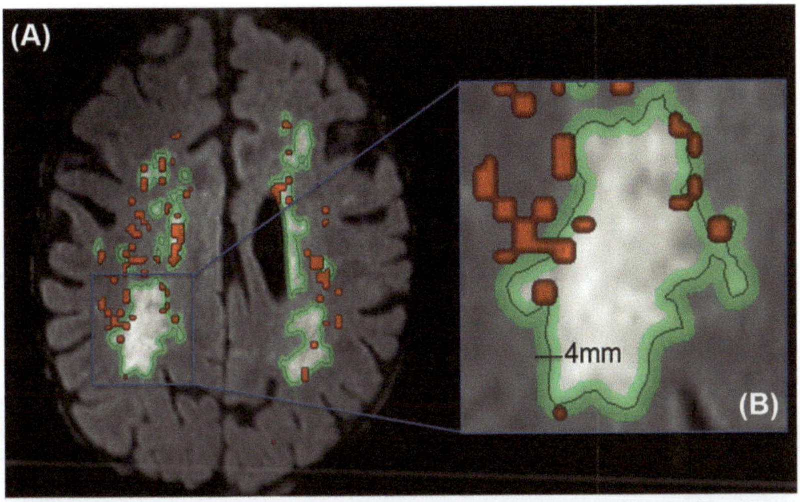

FIGURE 7.6 **Regions of increased permeability with DCE-MRI in Binswanger's disease.** WMH border and ring. (A) Fluid attenuated inversion recovery MRI shows the WMHs outlined with a green band to show the construction of the rim around the WMHs. (B) Higher magnification of one of the traced WMHs shows the border of the WMH (*black line*) and the 2 mm on both sides of the WMH border (*green rim*). Red areas show the increased permeability in the normal-appearing white matter. White area is WMH in FLAIR scan. *DCE*, Dynamic contrast-enhanced; *FLAIR*, fluid attenuated inversion recovery; *MRI*, magnetic resonance imaging; *WMHs*, white matter hyperintensities. *From Huisa, B. N., Caprihan, A., Thompson, J., Prestopnik, J., Qualls, C. R., & Rosenberg, G. A. (2015). Long-term blood-brain barrier permeability changes in Binswanger disease.* Stroke, *46(9), 2413–2418. https://doi.org/10.1161/STROKEAHA.115.009589.*

(Fig. 7.6) (Huisa et al., 2015). The increased signal seen on FLAIR MRI in the white matter is often interpreted as due to ischemia. In fact, large areas of white matter-increased FLAIR signal can be seen in healthy elderly individuals and probably related to aging. In a study of 100 normal elderly, seven had extensive white matter hyperintensities on FLAIR without any evidence of ischemic injury (Hunt et al., 1989). The term "normal-appearing white matter" refers to regions lacking increased FLAIR signal that have evidence of increased permeability. This suggests that these regions are vulnerable to develop into actual regions of white matter hyperintensities, but this is not known. Aging impairs the BBB permeability as shown by both albumin index and DCE-MRI (Thrippleton et al., 2019; Verheggen et al., 2020).

References

Abbott, N. J., Bundgaard, M., & Cserr, H. F. (1985). Tightness of the blood-brain barrier and evidence for brain interstitial fluid flow in the cuttlefish, *Sepia officinalis*. *The Journal of Physiology*, *368*, 213–226.

Akiguchi, I., Tomimoto, H., Suenaga, T., Wakita, H., & Budka, H. (1998). Blood-brain barrier dysfunction in Binswanger's disease; an immunohistochemical study. *Acta Neuropathologica*, *95*(1), 78–84.

Bell, R. D., Winkler, E. A., Singh, I., Sagare, A. P., Deane, R., Wu, Z., Holtzman, D. M., Betsholtz, C., Armulik, A., Sallstrom, J., Berk, B. C., & Zlokovic, B. V. (2012). Apolipoprotein E controls cerebrovascular integrity via cyclophilin A. *Nature*, *485*(7399), 512–516. Available from https://doi.org/10.1038/nature11087.

Benjamin, P., Zeestraten, E., Lambert, C., Ster, I. C., Williams, O. A., Lawrence, A. J., Patel, B., MacKinnon, A. D., Barrick, T. R., & Markus, H. S. (2016). Progression of MRI markers in cerebral small vessel disease: Sample size considerations for clinical trials. *Journal of Cerebral Blood Flow and Metabolism: Official Journal of the International Society of Cerebral Blood Flow and Metabolism*, *36*(1), 228–240.

Blevins, B. L., Vinters, H. V., Love, S., Wilcock, D. M., Grinberg, L. T., Schneider, J. A., Kalaria, R. N., Katsumata, Y., Gold, B. T., Wang, D. J. J., Ma, S. J., Shade, L. M. P., Fardo, D. W., Hartz, A. M. S., Jicha, G. A., Nelson, K. B., Magaki, S. D., Schmitt, F. A., Teylan, M. A., & Nelson, P. T. (2021). Brain arteriolosclerosis. *Acta Neuropathologica*, *141*(1), 1–24. Available from https://doi.org/10.1007/s00401-020-02235-6.

Bundgaard, M., & Abbott, N. J. (2008). All vertebrates started out with a glial blood-brain barrier 4–500 million years ago. *Glia*, *56*(7), 699–708. Available from https://doi.org/10.1002/glia.20642.

Choi, Y. H., & Yu, A.-M. (2014). ABC transporters in multidrug resistance and pharmacokinetics, and strategies for drug development. *Current Pharmaceutical Design*, *20*(5), 793–807.

Ewing, J. R., Wei, L., Knight, R. A., Pawa, S., Nagaraja, T. N., Brusca, T., Divine, G. W., & Fenstermacher, J. D. (2003). Direct comparison of local cerebral blood flow rates measured by MRI arterial spin-tagging and quantitative autoradiography in a rat model of experimental cerebral ischemia. *Journal of Cerebral Blood Flow and Metabolism: Official Journal of the International Society of Cerebral Blood Flow and Metabolism*, *23*(2), 198–209.

Hawkins, B. T., & Davis, T. P. (2005). The blood-brain barrier/neurovascular unit in health and disease. *Pharmacological Reviews*, *57*(2), 173–185.

Huisa, B. N., Caprihan, A., Thompson, J., Prestopnik, J., Qualls, C. R., & Rosenberg, G. A. (2015). Long-term blood-brain barrier permeability changes in Binswanger disease. *Stroke; a Journal of Cerebral Circulation*, *46*(9), 2413–2418. Available from https://doi.org/10.1161/STROKEAHA.115.009589.

Hunt, A. L., Orrison, W. W., Yeo, R. A., Haaland, K. Y., Rhyne, R. L., Garry, P. J., & Rosenberg, G. A. (1989). Clinical significance of MRI white matter lesions in the elderly. *Neurology*, *39*(11), 1470–1474.

Knight, R. A., Karki, K., Ewing, J. R., Divine, G. W., Fenstermacher, J. D., Patlak, C. S., & Nagaraja, T. N. (2009). Estimating blood and brain concentrations and blood-to-brain influx by magnetic resonance imaging with step-down infusion of Gd-DTPA in focal transient cerebral ischemia and confirmation by quantitative autoradiography with Gd-[(14)C]DTPA. *Journal of Cerebral Blood Flow and Metabolism: Official Journal of the International Society of Cerebral Blood Flow and Metabolism*, *29*(5), 1048–1058. Available from https://doi.org/10.1038/jcbfm.2009.20.

Montagne, A., Barnes, S. R., Sweeney, M. D., Halliday, M. R., Sagare, A. P., Zhao, Z., Toga, A. W., Jacobs, R. E., Liu, C. Y., Amezcua, L., Harrington, M. G., Chui, H. C., Law, M., & Zlokovic, B. V. (2015). Blood-brain barrier breakdown in the aging human hippocampus. *Neuron*, *85*(2), 296–302. Available from https://doi.org/10.1016/j.neuron.2014.12.032.

Nag, S., Robertson, D. M., & Dinsdale, H. B. (1979). Quantitative estimate of pinocytosis in experimental acute hypertension. *Acta Neuropathologica*, *46*(1–2), 107–116.

Neltner, J. H., Abner, E. L., Baker, S., Schmitt, F. A., Kryscio, R. J., Jicha, G. A., Smith, C. D., Hammack, E., Kukull, W. A., Brenowitz, W. D., Van Eldik, L. J., & Nelson, P. T. (2014). Arteriolosclerosis that affects multiple brain regions is linked to hippocampal sclerosis of ageing. *Brain : A Journal of Neurology*, *137*(Pt 1), 255–267. Available from https://doi.org/10.1093/brain/awt318.

Nigro, P., Pompilio, G., & Capogrossi, M. C. (2013). Cyclophilin A: A key player for human disease. *Cell Death & Disease*, *4*, e888. Available from https://doi.org/10.1038/cddis.2013.410.

Ohno, K., Pettigrew, K. D., & Rapoport, S. I. (1978). Lower limits of cerebrovascular permeability to nonelectrolytes in the conscious rat. *The American Journal of Physiology*, *235*(3), H299–H307.

Patlak, C. S., & Blasberg, R. G. (1985). Graphical evaluation of blood-to-brain transfer constants from multiple-time uptake data. Generalizations. *Journal of Cerebral Blood Flow and Metabolism : Official Journal of the International Society of Cerebral Blood Flow and Metabolism*, *5*(4), 584–590.

Raichle, M. E., Eichling, J. O., & Grubb, R. L. (1974). Brain permeability of water. *Archives of Neurology*, *30*(4), 319–321.

Raja, R., Rosenberg, G. A., & Caprihan, A. (2018). MRI measurements of blood-brain barrier function in dementia: A review of recent studies. *Neuropharmacology*, *134*(Pt B), 259–271. Available from https://doi.org/10.1016/j.neuropharm.2017.10.034.

Rasmussen, M. K., Mestre, H., & Nedergaard, M. (2018). The glymphatic pathway in neurological disorders. *The Lancet. Neurology*, *17*(11), 1016–1024. Available from https://doi.org/10.1016/S1474-4422(18)30318-1.

Rocha, E. M., De Miranda, B., & Sanders, L. H. (2018). Alpha-synuclein: Pathology, mitochondrial dysfunction and neuroinflammation in Parkinson's disease. *Neurobiology of Disease*, *109*(Pt B), 249–257. Available from https://doi.org/10.1016/j.nbd.2017.04.004.

Román, G. C. (1987). Senile dementia of the Binswanger type. A vascular form of dementia in the elderly. *JAMA: the Journal of the American Medical Association*, *258*(13), 1782–1788.

Rosenberg, G. A., Kyner, W. T., & Estrada, E. (1980). Bulk flow of brain interstitial fluid under normal and hyperosmolar conditions. *The American Journal of Physiology*, *238*(1), F42–F49.

Smith, E. E., Biessels, G. J., De Guio, F., de Leeuw, F. E., Duchesne, S., Düring, M., Frayne, R., Ikram, M. A., Jouvent, E., MacIntosh, B. J., Thrippleton, M. J., Vernooij, M. W., Adams, H., Backes, W. H., Ballerini, L., Black, S. E., Chen, C., Corriveau, R., DeCarli, C., & Wardlaw, J. M. (2019). Harmonizing brain magnetic resonance imaging methods for vascular contributions to neurodegeneration. *Alzheimer's & Dementia (Amsterdam, Netherlands)*, *11*, 191–204. Available from https://doi.org/10.1016/j.dadm.2019.01.002.

Sonnen, J. A., Santa Cruz, K., Hemmy, L. S., Woltjer, R., Leverenz, J. B., Montine, K. S., Jack, C. R., Kaye, J., Lim, K., Larson, E. B., White, L., & Montine, T. J. (2011). Ecology of the aging human brain. *Archives of Neurology*, *68*(8), 1049–1056. Available from https://doi.org/10.1001/archneurol.2011.157.

Srivastava, T., Sherman, L. S., & Back, S. A. (2020). Dysregulation of hyaluronan homeostasis during white matter injury. *Neurochemical Research*, *45*(3), 672–683. Available from https://doi.org/10.1007/s11064-019-02879-1.

Sweeney, M. D., Zhao, Z., Montagne, A., Nelson, A. R., & Zlokovic, B. V. (2019). Blood-brain barrier: From physiology to disease and back. *Physiological Reviews*, *99*(1), 21–78. Available from https://doi.org/10.1152/physrev.00050.2017.

Taheri, S., Gasparovic, C., Huisa, B. N., Adair, J. C., Edmonds, E., Prestopnik, J., Grossetete, M., Shah, N. J., Wills, J., Qualls, C., & Rosenberg, G. A. (2011). Blood-brain barrier permeability abnormalities in vascular cognitive impairment. *Stroke; a Journal of Cerebral Circulation*, *42* (8), 2158–2163. Available from https://doi.org/10.1161/STROKEAHA.110.611731.

Tam, S. J., & Watts, R. J. (2010). Connecting vascular and nervous system development: Angiogenesis and the blood-brain barrier. *Annual Review of Neuroscience*, *33*, 379–408. Available from https://doi.org/10.1146/annurev-neuro-060909-152829.

Thrippleton, M. J., Backes, W. H., Sourbron, S., Ingrisch, M., van Osch, M. J. P., Dichgans, M., Fazekas, F., Ropele, S., Frayne, R., van Oostenbrugge, R. J., Smith, E. E., & Wardlaw, J. M. (2019). Quantifying blood-brain barrier leakage in small vessel disease: Review and consensus recommendations. *Alzheimer's & Dementia : The Journal of the Alzheimer's Association*, *15*(6), 840–858. Available from https://doi.org/10.1016/j.jalz.2019.01.013.

Tourtellotte, W. W., Staugaitis, S. M., Walsh, M. J., Shapshak, P., Baumhefner, R. W., Potvin, A. R., & Syndulko, K. (1985). The basis of intra-blood-brain-barrier IgG synthesis. *Annals of Neurology*, *17*(1), 21–27.

van de Haar, H. J., Burgmans, S., Jansen, J. F. A., van Osch, M. J. P., van Buchem, M. A., Muller, M., Hofman, P. A. M., Verhey, F. R. J., & Backes, W. H. (2016). Blood-brain barrier leakage in patients with early Alzheimer disease. *Radiology*, *281*(2), 527–535.

Venegas, C., Kumar, S., Franklin, B. S., Dierkes, T., Brinkschulte, R., Tejera, D., Vieira-Saecker, A., Schwartz, S., Santarelli, F., Kummer, M. P., Griep, A., Gelpi, E., Beilharz, M., Riedel, D., Golenbock, D. T., Geyer, M., Walter, J., Latz, E., & Heneka, M. T. (2017). Microglia-derived ASC specks cross-seed amyloid-β in Alzheimer's disease. *Nature*, *552* (7685), 355–361. Available from https://doi.org/10.1038/nature25158.

Verheggen, I. C. M., de Jong, J. J. A., van Boxtel, M. P. J., Gronenschild, E. H. B. M., Palm, W. M., Postma, A. A., Jansen, J. F. A., Verhey, F. R. J., & Backes, W. H. (2020). Increase in blood-brain barrier leakage in healthy, older adults. *GeroScience*, *42*(4), 1183–1193. Available from https://doi.org/10.1007/s11357-020-00211-2.

Wallin, A., Blennow, K., Fredman, P., Gottfries, C. G., Karlsson, I., & Svennerholm, L. (1990). Blood brain barrier function in vascular dementia. *Acta Neurologica Scandinavica*, *81*(4), 318–322.

Wang, X., Michaelis, M. L., & Michaelis, E. K. (2010). Functional genomics of brain aging and Alzheimer's disease: Focus on selective neuronal vulnerability. *Current Genomics*, *11*(8), 618–633. Available from https://doi.org/10.2174/138920210793360943.

Wong, S. M., Jansen, J. F. A., Zhang, C. E., Hoff, E. I., Staals, J., van Oostenbrugge, R. J., & Backes, W. H. (2019). Blood-brain barrier impairment and hypoperfusion are linked in cerebral small vessel disease. *Neurology*, *92*(15), e1669–e1677. Available from https://doi.org/10.1212/WNL.0000000000007263.

Chapter 8

Mechanisms of white matter injury

White matter is the major site of pathology in vascular-mediated dementia

The amyloid hypothesis of Alzheimer's disease (AD) has dominated research in dementia since the discovery of amyloid in the plaques first described by Alzheimer in a patient with early-onset dementia (Glenner, 1981). The dominance of the amyloid hypothesis limited research money available to study other illnesses that caused dementia. In addition, the relative lack of interest in diseases of aging led to fewer autopsies being performed on those dying of old age, and a lack of information on injury to the white matter subsequent to the pathological findings in the white matter described in eight patients by Otto Binswanger, a contemporary of Alzheimer. Another factor in the movement away from the amyloid hypothesis was the failure of several large clinical trials of antiamyloid agents that successfully removed amyloid from the brain without improving cognition. Magnetic resonance imaging (MRI) revived interest in white matter by showing the extent of white matter changes in patients with dementia. This has redirected studies in dementia to other causes of dementia, particularly those related to the white matter. This renewed interest in white matter disease has complimented the early emphasis on gray matter (GM), where the primary damage of AD occurs to the cortx and hippocampus. In conjunction with the growth of neuroimaging, there has been a dramatic increase in the understanding of the molecular events that lead to white matter injury. The growth in the use of biomarkers in diagnosis and treatment trial design has been another factor in the increased interest in understanding the role of white matter in cognitive diseases. There has been a significant increase in the number of studies designed to unravel the molecular basis for white matter injury, which have led to improved methods of classifying patients with dementia. Biomarkers obtained from neuroimaging and biochemical measurements in blood and CSF have played a major role in this effort.

Neuroinflammation in Vascular Dementia. DOI: https://doi.org/10.1016/B978-0-12-823455-6.00003-1

Biomarkers in patient classification

As the limitations of using the changes in the MRI as indicators of disease became clearer, particularly in patients with multiple pathological processes those with MRI changes of aging that confounded diagnosis, the need emerged for a multimodal approach to diagnosis to aid in separating the overlapping diagnoses in a heterogeneous group of patients with abnormalities in the white matter observed on MRI. The multimodal approach utilizes information obtained from neuropsychological testing, MRI, and the biochemical analysis of blood and CSF to arrive at a more accurate diagnosis. With the growth of information available for diagnosis of an individual patient, the ability to use conventional methods to analyze the data has become inadequate, creating the need for methods that are used for the analysis of "big data," which involves computer-aided artificial intelligence. Neurodegenerative diseases of the GM have been well described pathologically, including Alzheimer's disease, frontotemporal dementia (FTD), Parkinson's disease with dementia, and others. Diseases of the white matter have not seen similar progress. As opposed to GM where neurons are the major site if pathological change, the vulnerability of the white matter does not reside in a single cell type, but rather involves a collection of tissue extending from below the cortex to the edge of the ventricles; the vulnerability is based on the structure of the brain vasculature and the involvement of disorders of the circulation. Arterioles are the primary blood vessels that are damaged, and the main causes of the injury are hypertension and diabetes.

The major function of white matter is transmission of electrical signals along fiber tracts, delivery of nutrients, and the removal of metabolic waste products. Axons protruding from the cell bodies conduct the electrical signals in a slow fashion unless they are covered with myelin, which increases the speed of nerve propagation many fold by saltatory conduction from one node of Ranvier to another. Diseases of myelin fall into two types: those with failure in the formation of myelin, which mainly manifest in childhood. Myelin disorders of childhood lead to poorly myelinated fibers in hereditary and metabolic diseases. By adolescence, the myelination process has almost completed and the disorders of myelin are related to destruction of myelin by infection, inflammation, and injury. Myelin is made by oligodendrocytes, which are sensitive to hypoxia, and injury to these cells leads to loss of myelin. Aging has detrimental effects on myelin that are subtle and detectable by MRI. Cerebral blood flow to the white matter is much lower than that in GM, making the white matter vulnerable to reductions in blood flow and low oxygen. Furthermore, white matter can be damaged by toxins, such as solvents and torn by trauma, particularly when the injury causes a sudden twisting motion. Diseases that interfere with the normal flow of interstitial fluid can lead to accumulation of toxic molecules that damage white matter.

Injured regions of white matter can be readily visualized with appropriate MRI pulse sequences. Fluid attenuated inversion recovery (FLAIR) is a MRI pulse sequence that nulls the water signal from unrestrained, freely moving water that is mainly in the ventricles, emphasizing tissue water. This sequence is excellent for demonstrating edema fluid in the white matter, and the accumulation of water around the ventricles in normal pressure hydrocephalus where the removal of water from the brain is blocked and the water moves by transependymal flow into the periventricular tissue. Because of the multiple reasons for increase in the FLAIR signal, it is of limited use in diagnosis for individual patients, but can provide important information in studies of large populations. Aging is a confounding factor in interpretation of MRI FLAIR signal since increases in the FLAIR signal are commonly found in healthy elderly. Several MRI methods are more sensitive to damage to white matter, and these can be used to separate an injury signal from normal aging. Of these methods, the two most useful are proton magnetic resonance spectroscopy (^{1}H-MRS) and diffusion tensor imaging (DTI) both of which show white matter damage. The spectra obtained from ^{1}H-MRS have a large signal from *N*-acetylaspartate (NAA) because of its high concentration in brain; injury to the white matter results in a loss of the NAA signal. In spite of abundance in brain, its role is not well understood. Another MRI method to show injury is DTI; changes in the diffusion characteristics of water are an excellent indicator of injury to the white matter. While ^{1}H-MRS is more complicated to implement and to interpret, the DTI signal can be obtained in the routine clinical MRI scanner, making it the preferred method to show regions of white matter injury. because it is available on most MRI instruments, and does not require special MRI expertise. However, either can be used since NAA and DTI fractional anisotropy (FA) are highly correlated. One study analyzed a range of conditions from normotension, through hypertension, to symptomatic cerebral small vessel disease (SVD); they found an increase in mean diffusivity (MD) and a reduction in FA in the SVD cohort. There was a consistent and robust relationship between these two DTI parameters and NAA, a marker of neuronal injury. These correlations were present in all three groups. Axonal loss/dysfunction is a major pathological process underlying the reduction in MD and increased FA seen in both symptomatic and asymptomatic cerebral SVD. The strength of the correlations increased with disease severity and was greatest in symptomatic subjects. This finding reflects the greater range of FA and MD values seen in the symptomatic SVD cohort: a relationship that was independent of white matter lesion volume. The correlation between NAA and FA was weaker than that with MD—possibly due to FA being a more heterogenous white matter measure than MD (Nitkunan et al., 2008).

Although AD is primarily a disease of the cortex, there is evidence from both pathological studies and imaging to show involvement of the white matter. Selective incomplete white matter infarctions were frequently seen in dementia of the Alzheimer's type: brain tissue showed white matter with incomplete infarction, complete infarcts and histologically normal tissue. In cases with dementia, mainly of Alzheimer and multi-infarct type, in non-demented subjects with cerebral infarcts and in age-matched control cases, the etiological significance of the loss of different white matter components supported the concept of selective incomplete white matter infarctions as being an independent white matter disorder of cerebrovascular, hypoperfusional/hypoxic origin (Englund et al., 1988).

In an MRI study of patients with AD, there was both WMH and entorhinal cortex volume changes that were predictive of rapid progression in early stages of the disease and they interacted synergistically. The findings implicate small vessel cerebrovascular disease in AD pathogenesis and clinical expression. The presence of SVD in the brains of AD patients suggested avenues for prevention or treatment of rapid progressing course among patients with MCI (Tosto et al., 2014). There is growing evidence to support the importance of the confluence of AD-type pathology and cerebrovascular disease as an indicator of a more aggressive form of dementia.

Myelin disruption in inflammation

Astrocytes, oligodendrocytes, and microglia are the main cells of the nervous system. In white matter, the oligodendrocytes produce the myelin that covers the axons, facilitating transmission of electrical signals from neurons. One oligodendrocyte is able to provide myelin for many axons, leading to widespread damage if they are injured. Astrocytes are support cells; they take up glutamate to protect from excitotoxic damage. Glutamine synthetase converts the glutamate into glutamine. Following injury, the astrocytes cause fibrosis with the production of glial fibrillary acidic protein (GFAP). Extensive fibrosis interferes with function of the white matter and blocks regenerating fibers during repair. Microglia are immune cells that act as macrophages for the brain: they enter the brain from the bone marrow during the early stages of development. White matter has a large number of microglia that are inactive prior to an injury. Oligodendrocytes are embedded in a vast network of interconnected glial and neuronal cells, in which oligodendrocytes actively provide metabolic support to neurons, regulate ion and water homeostasis, and adapt to activity-dependent neuronal signals. Oligodendrocytes take up glucose to form lactate and pyruvate, which is delivered to the axons using monocarboxylate transporters, providing the axons with energy. Neuronal activity results in the release of glutamate into the periaxonal space, where it activates *N*-methyl-D-aspartate (NMDA) receptors at the inner myelin membrane. This in turn triggers the translocation of glucose transporters (GLUT1) into the surface of the myelin sheath and/or oligodendrocyte cell body to increase glucose uptake and the availability of lactate/pyruvate. Astrocytes are connected to oligodendrocytes by gap junctions to enable metabolite transfer between the cells. Specific interactions between astrocytes, oligodendrocytes, and microglia contribute to demyelination in hereditary white matter pathologies, and in inflammatory demyelinating diseases, such as multiple sclerosis, which is increasingly recognized as a "whole-brain" disease (Stadelmann et al., 2019).

Glutamate-mediated excitotoxicity

In GM, the principal mechanism of acute ischemic injury is excitotoxicity involving glutamate release, glutamate receptor (GluR) activation, cytotoxic Ca^{2+} elevation, and further glutamate release. A similar mechanisms is involved in myelin damage in hypoxic injury; glutamate causes excitotoxic injury mediated by myelinic GluR over-activation. In addition to myelin, white matter GluRs are expressed on the oligodendrocyte cell bodies responsible for generating and maintaining myelin, and these cell bodies are also highly susceptible to excitotoxic ischemic injury (Fern & Matute, 2019).

Glutamate's excitotoxicity is based on the ionotropic and metabotropic receptors that open calcium and sodium channels; if the glutamate is not rapidly removed from the synapse, the influx of calcium activates molecular processes that lead to cell death. There are several types of GluRs: NMDA are on the oligodendrocyte cell body and the α-amino-3-hydrocy-5-methyl-4-isoxazole propionic acid (AMPA)/kainate (KA) receptors are on the axons. Uptake of glutamate is facilitated by astrocytes; glutamate cannot cross membranes, but it is converted to glutamine and taken up into other cells to complete the recycling process. When there is a reduction of energy, glutamate remains elevated, leading to further injury.

Oxygen and nitrogen free radicals are both necessary for many normal functions and contribute to pathological processes. Reactive oxygen species (ROS) can activate enzymes in the matrix metalloproteinase (MMP) family. Hydrogen peroxide (H_2O_2) is formed from an oxygen radical in the presence of superoxide dismutase (SOD), which is further metabolized by glutathione to water and oxygen. This natural antioxidant is present in all cells and provides important protective functions. However, when excessive amounts of peroxidase are formed in the presence of iron, the Fenton Reaction converts the peroxidase to toxic oxygen radicals that injure cells.

Nitric oxide (NO), the smallest signaling molecule known, is produced by three isoforms of NO synthase. They all utilize l-arginine and molecular oxygen as substrates and require the cofactors reduced nicotinamide-adenine-dinucleotide phosphate (NADPH), flavin adenine dinucleotide (FAD), flavin mononucleotide (FMN), and (6*R*-)5,6,7,8-tetrahydrobiopterin (BH4). Neuronal NOS (nNOS) is constitutively expressed in central and peripheral neurons and some other cell types. Inducible NOS (iNOS) can be expressed in many cell types in response to lipopolysaccharide, cytokines, or other agents. Inducible NOS generates large amounts of NO that contributes to the pathophysiology of inflammatory diseases. Inducible NOS is not usually expressed in cells, but its expression can be induced by bacterial lipopolysaccharide, cytokines, and other agents. Although primarily identified in macrophages, expression of the enzyme can be stimulated in virtually any cell or tissue, provided that the appropriate inducing agents have been identified. Endothelial NOS (eNOS) is expressed in endothelial cells. It keeps blood vessels dilated, controls blood pressure, and has numerous other vasoprotective and antiatherosclerotic effects (Förstermann & Sessa, 2012).

Inflammation in small vessel disease of the Binswanger's type

Multiple lines of evidence indicate that inflammation is important in the pathophysiology of Binswanger's disease; in addition to that from disruption of the blood–brain barrier (BBB) in the white matter, there is evidence of microglia activity in the white matter accompanied by increased MMP expression. The brains of patients with Binswanger's disease have an absence of amyloid and phosphorylated tau, but have activated microglia and GFAP-positive astrocytes in the white matter. The characteristic findings are extensive loss of myelin extending into the sulci, sparing of the U-fibers, and fibrosis of the white matter with GFAP immunostaining (Fig. 8.1). The U-fibers are the association fibers that connect regions of the cortex. In typical Binswanger's disease with extensive loss of myelin in the central regions,

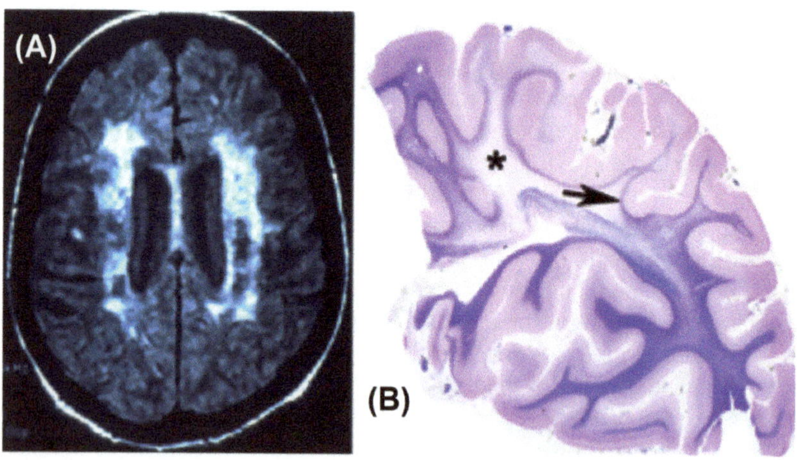

FIGURE 8.1 **Myelin loss in regions of white matter hyperintensities.** (A) FLAIR MRI in Binswanger's patient. The large white matter hyperintensities are seen without atrophy. (B) Myelin staining in a different patient with Binswanger's disease shows the demyelination in the centrum semiovale (*asterisk*). The area shows the sparing of the U-fibers, supporting the role of inflammation in the damage to the white matter. *MRI*, magnetic resonance imaging, *FLAIR*, Fluid attenuated inversion recovery. *From Rosenberg, G. A., Sullivan, N., & Esiri, M. M. (2001). White matter damage is associated with matrix metalloproteinases in vascular dementia.* Stroke, 32*(5), 1162–1168.*

the U-fibers remain intact. This is strong evidence that the pathological process is one of slowly progressive inflammation rather than a series of strokes.

Regions of demyelination appear as white matter hyperintensities in FLAIR MRI; they lack myelin staining except in the U-fibers that join association areas; there is BBB disruption around the white matter hyperintensities (Huisa et al., 2015; Rosenberg, 2017). Microglia cells are increased particularly in the regions with loss of myelin (Rosenberg et al., 2001) (Fig. 8.2).

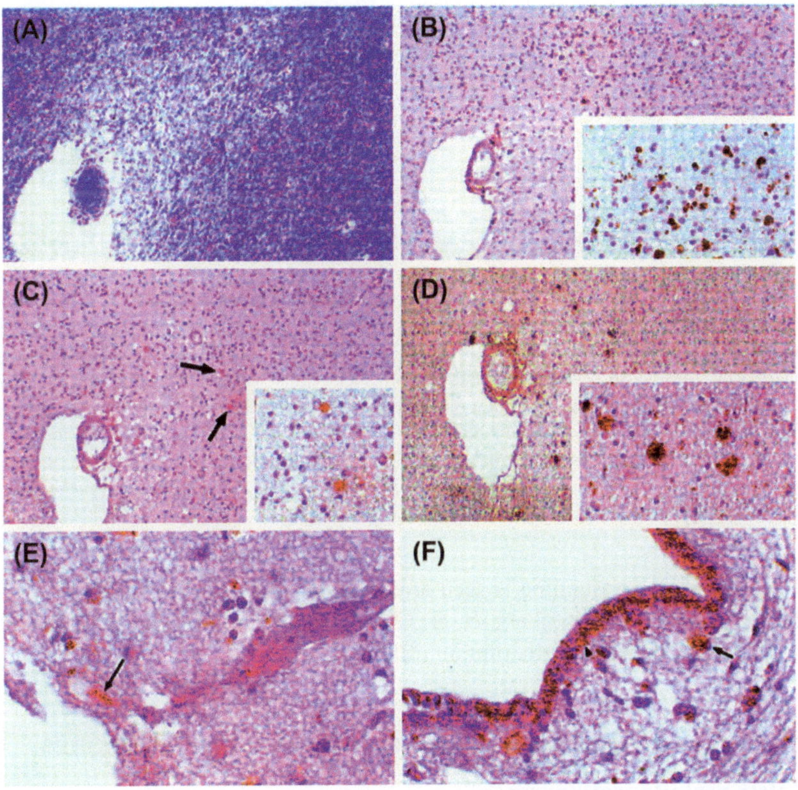

FIGURE 8.2 Small demyelinated region in a control brain. (A) Myelin stain showing blood vessel with enlarged perivascular space and inflammatory cells near the region of demyelination (original magnification ×100). (B) Same region as in panel A (original magnification ×100). Insert is higher power of demyelinated region and indicates the presence of PG-M1+ cells (original magnification ×400). (C) MMP-2+ cells (arrows) in a low-power field (original magnification ×100) with insert showing several PG-M1+ cells (original magnification ×400). (D) Same area as in panel A showing MMP-3 staining in cells in the same area (original magnification ×100). Insert is higher power (original magnification ×400). (E) Subpial zone with MMP-2 reactivity in cells near penetrating blood vessel (*arrow*) (original magnification ×400). (F) Ependymal cells with MMP-2+ cells (*arrows*) beneath them and fine processes extending through them into the ventricle (original magnification ×400). *MMP*, Matrix metalloproteinase. *From Rosenberg, G. A., Sullivan, N. & Esiri, M. M. (2001). White matter damage is associated with matrix metalloproteinases in vascular dementia.* Stroke, *32(5), 1162–1168.*

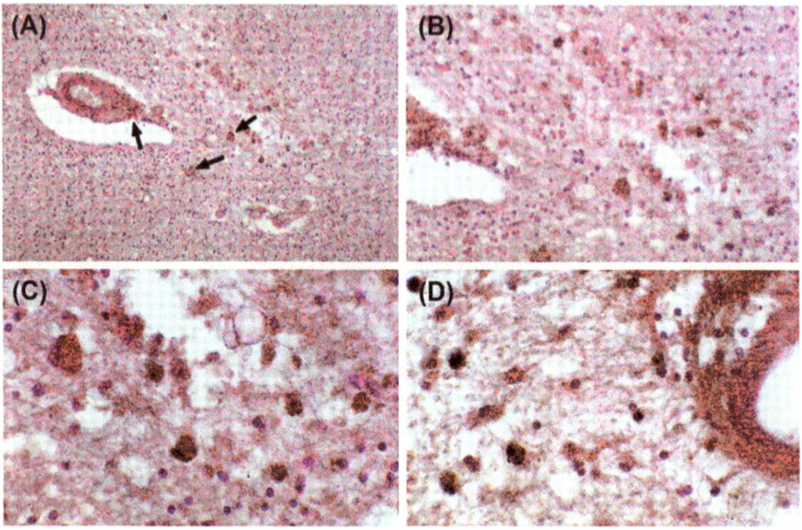

FIGURE 8.3 MMP-3–containing cells in Binswanger's disease (BD) subject. (A) Arteriosclerotic vessel with fibrohyaline changes. MMP-3 + cells are seen surrounding the vessel and in the adjacent parenchyma (*arrows*) (original magnification ×100). (B) Large tissue macrophages with MMP-3 in severely affected white matter region (original magnification ×200). (C) Area of demyelination distant from a blood vessel with intense MMP-3 + cells (original magnification ×400). (D) Area closer to a blood vessel showing many MMP-3 + tissue macrophages (original magnification ×400). *MMP*, Matrix metalloproteinase. *From Rosenberg, G. A., Sullivan, N., & Esiri, M. M. (2001). White matter damage is associated with matrix metalloproteinases in vascular dementia.* Stroke, *32(5), 1162–1168.*

MMP-3 immunostaining is seen around blood vessels in microglia/macrophages (Fig. 8.3) (Rosenberg et al., 2001). MMP-3 is in the pericytes as shown in this double-labeled immunofluorescent image.

Interaction of proteases, angiogenic factors, and cytokines

Matrix metalloproteinases

Myelin breakdown occurs secondary to a number of disease processes that can directly or indirectly attack the myelinated fibers. Many cell types that are activated in inflammation secrete substances that participate in myelin damage. Proteases are released by a number of inflammatory cells, particularly the endogenous microglia and the invading macrophages, but also astrocytes and endothelial cells. Free radicals and proteases are the major agents involved in myelin damage. The first studies to demonstrate the destructive power of proteases were done with plasmin, a ubiquitous enzyme released in normal processes such as removal of clots, but which can activate a number of other proteases, creating cascades of toxic products. Neutral proteases, such as plasminogen activator (PA), released from activated macrophages damage myelin. The damage to the myelin only occurs in the presence of a complement. Aminomethylcyclohexane carboxylic acid (AMCA) and other neutral protease inhibitors protected myelin in animals with experimental allergic encephalomyelitis (Brosnan et al., 1980). MMPs are a family of proteases that attack all components of the extracellular matrix; they degrade myelinated fibers and cause demyelination (Rosenberg, 2017). The two major systems that modify the extracellular matrix in brain are the PA and MMP axes. Deleterious effects include the disruption of BBB integrity, amplification of inflammatory infiltrates, demyelination, and possibly interruption of cell-cell and cell-matrix interactions that may trigger cell death. In contrast, PA-MMP actions may contribute to extracellular proteolysis that mediates parenchymal and angiogenic recovery after brain injury (Lo et al., 2002). MMPs that degrade myelin include: interstitial collagenase (MMP-1), 72 kDa gelatinase (MMP-2), stromelysin-1 (MMP-3), matrilysin (MMP-7) and 92 kDa gelatinase (MMP-9); they all were able to digest myelin basic protein (MBP), the major extrinsic membrane protein of central nervous system myelin. MMP-2 had the greatest impact on myelin, followed by MMP-3, MMP-1, MMP-7, and MMP-9, which had comparable effects, but with less activity. Production of these enzymes by glia or infiltrating inflammatory cells could therefore contribute to demyelination in neuroinflammatory disease (Chandler et al., 1995). Demyelination during an inflammatory response that involves microglia or macrophages, both of which release proteases, can attack the myelinated fibers, breaking down myelin. These same MMPs are involved in disruption of the BBB, these are two MMP-mediated pathological processes form the pathological basis for the white matter damage in Binswanger's disease.

Angiogenesis

After an ischemic or traumatic injury, blood vessels begin the process of regrowing to aid recovery by increasing cerebral blood flow to the injured regions. The growth of new capillaries from pre-existing blood vessels is mediated by several growth factors in the family of cytokines called vascular endothelial growth factor (VEGF). The VEGF family is composed of five members: VEGF-A (often referred to as VEGF), VEGF-B, VEGF-C, VEGF-D, and placental growth factor (PlGF). Angiogenesis is common in tumors where they increase the blood supply to maintain tissue oxygenation in the growing hypoxic tumors. Drugs that block angiogenesis have been used to slow the growth of tumors, but with only limited success (Folkman, 2002). Hypoxia initiates a similar process as the hypoxic tissue releases hypoxia inducible factor-1α (HIF-1α) in order to increase the delivery of blood to the injured tissue. Although angiogenesis is a well-described mechanism post injury, it remains unclear whether it is detrimental or beneficial. The main molecule responsible for regrowth of vessels is VEGF, which acts through receptors on the blood vessels, flt-1 and tie-2; these two isoforms bind VEGF receptors, VEGFR1 (fms-related tyrosine kinase 1 or Flt-1 in humans) and VEGFR2 (kinase insert domain receptor also designated as Flk-1, or KDR in humans). PlGF is another molecule that participates in the regrowth of blood vessels. Hypoxia and reactive oxygen species are strong inducers of VEGF family members, including PlGF. PLGF is increased in CSF of FTD. Expression of PlGF is increased in mouse astrocytes and endothelial cells following cerebral ischemia. Astrogliosis in frontal and temporal regions is one of the core histopathological hallmarks of FTD. Furthermore, frontotemporal lobar degeneration has been shown to be accompanied by oxidative damage that targets primarily astrocytes. Thus it is possible that in FTD, PlGF is increased in response to astrogliosis and oxidative stress (Hansson et al., 2019).

Angiogenic vessels differ from patent vessels in that they have numerous fenestrae, widened inter-endothelial junctions, abnormal endothelial cell shapes, and abnormal basement membranes. Angiogenic vessels can be distinguished from normal vessels by using antibodies to integrin, $\alpha\nu\beta3$, a dimeric glycoprotein expressed on the endothelial cell surface. Integrin $\alpha\beta\nu3$ is absent on patent vessels, but dramatically increased on angiogenic vessels; there was an increase in $\alpha\nu\beta3$ immunoreactivity and vessel number in the hippocampus, one of the first areas to show pathologic changes indicative of AD (Desai et al., 2009).

Studies of CSF in patients with dementia have shown an increase in both VEGF and transforming growth factor-beta (TGF-β), suggesting that they may be involved in the disruption of the BBB. Patients with AD and vascular dementia have intrathecal production of VEGF, a factor which plays a pivotal role in angiogenesis, supporting the idea that vascular factors might play a role in the pathogenesis of both these conditions. In addition, both dementia groups have intrathecal production of TGF-β, a cytokine that also has angiogenic properties (Tarkowski et al., n.d.).

The ratio of albumin in the CSF to that in the plasma (albumin index) indicates permeability of the BBB (Eeg-Olofsson et al., 1981). The albumin index was increased in diabetes mellitus and correlated positively with CSF biomarkers of angiogenesis and endothelial dysfunction (VEGF, intracellular adhesion molecule 1, and vascular cell adhesion molecule 1). In healthy elderly, high body mass index and waist-hip ratio predicted increased albumin index 20 years later. BBB permeability is increased in major dementia disorders but does not relate to amyloid pathology or APOE genotype. Instead, BBB impairment may be associated with diabetes and brain microvascular damage (Janelidze et al., 2017).

Corroboration of the importance of angiogenesis in white matter damage comes from studies in a model of vascular dementia in the hypertensive rat; these rats have chronic hypoxia with progressive damage to the white matter and opening of the BBB. Tight junction proteins, which maintain BBB integrity, are reduced in acute ischemia. In a study in spontaneously hypertensive/stroke prone rats with unilateral carotid artery occlusion and Japanese permissive diet (high salt and low protein), which is a model of white matter damage, there was tight junction protein degradation and angiogenesis associated with pathological progression of white matter damage as shown by MRI and IgG immunostaining that demonstrated regions with BBB damage that corresponded with decreased endothelial tight junction proteins, claudin-5, occludin, and ZO-1. Affected white matter had increased expression of angiogenic factors, Ki_{67}, NG2, VEGF-A, and MMP-3 in vascular endothelial cells and pericytes. Treatment with minocycline to block BBB disruption reduced WM lesion size, and extended survival, permitting the study of longer term effects. Minocycline-treated rats showed increased VEGF-A protein, tight junction proteins formation, and oligodendrocyte proliferation. The study showed that chronic hypoxia disrupts tight junction proteins, increasing vascular permeability, and initiates angiogenesis in white matter (Yang et al., 2018).

The mechanism of angiogenesis involves sprouting of the newly formed vessels off of existing blood vessels. The movement of the new vessels into the surrounding tissue involves formation of tip cells secondary to VEGF binding to VEGFR2, the VEGF receptor, during hypoxia, guiding new vascular branching. This process is mediated by astrocytes. Angiogenic/hypoxic markers accumulate around amyloid plaques, but paradoxically cerebral microvasculature is decreased, suggesting an abnormal angiogenic process in AD. Rather than producing normal functioning blood vessels, the initiation of angiogenesis around AD plaques leads to disassembly of amyloid-plaque and increased phagocytic activity of microglia. This results in vascular scars rather than functional blood vessels, which explains the reduced number of blood vessels around amyloid plaques (Alvarez-Vergara et al., 2021). Therefore the finding of angiogenic molecules, such as VEGF in the CSF of patients with AD, may not represent the increased production of functioning blood vessels.

Cytokines

Studies of biomarkers in vascular dementia show a combination of factors that provide clues to the underlying pathological changes. Although these studies are rarely longitudinal, they have a number of biomarkers that are commonly found in CSF, including MMP-2, MMP-3, MMP-9, and MMP-10; angiogenic growth factors such as VEGF, Flt-1, tie-2, and PlGF, and antiinflammatory cytokines. Other factors participate in vessel growth after injury, including chemokines, particularly chemokine (C-C motif) ligand 2 (CCL), neurofilament light, and inflammatory exosomes. Several of the biomarkers have been linked together in studies. Angiogenesis requires the coordinated effort of proteases to disrupt the extracellular matrix and to strip pericytes from the vessels: this allows for molecules, such as VEGF to initiate sprouting. Growth of vessels from scratch or vasculogenesis only occurs in early development. The VEGF receptors, Flt-1 and PlGF, are also induced. Neurofilament light is a marker of axonal injury and appears in the CSF in a number of neurological disorders. Likewise the cytokines, IL-1, IL-10, and IL-13, which are antiinflammatory, suggest that there is an attempt to heal the brain. If that is the case, it would be unwise to attempt to block angiogenesis or antiinflammatory cytokines, which will greatly complicate the search for effective treatments.

Bystander demyelination

Myelin breakdown occurs through the action of proteases, which are secreted by activated intrinsic microglia and recruited macrophages. Caspases and MMPs break down myelin, leaving the axon to function in an impaired fashion with slow conduction due to the loss of myelin. When there is an inflammatory response, proteases are released as an unintended consequence of the secreted enzymes; this is referred to as "by-stander" demyelination (Fig. 8.4). Since hypoxia causes activation of microglia cells and activation of metalloproteinases, there is the potential for the myelin to be damaged. Experimental autoimmune encephalomyelitis (EAE) is an inflammatory demyelinating disease of the central nervous system commonly used as a model for multiple sclerosis. In multiple sclerosis, demyelination occurs in association with perivascular infiltration of T-cells and macrophages. The similarities in immunopathology suggest that these two diseases share common mechanisms of tissue destruction. A general mechanism has been proposed to explain the clinical and histopathological features of EAE: T-cells sensitized to the inducing antigen, MBP, react with antigen-presenting cells (possibly endothelial cells, microglia or astrocytes) in the central nervous system. As a consequence of this reaction, T-cells release lymphokines which activate macrophages, augmenting an inflammatory response, leading to vasospasm and breakdown of the BBB. The activated macrophages secrete inflammatory mediators, including PA and other proteinases, which, in concert with serum plasminogen and complement, initiate myelin destruction (Norton et al., 1978, 1990).

Demyelination is found in cases of AD where in addition to the GM degeneration, a lesion that has the character of an incomplete infarction is observed in white matter. Sixty percent of early and late AD show pathological changes in the white matter with mild changes in two-thirds and moderate or severe changes in one-third. It involves the deep white matter symmetrically, tapering off toward the cortex. It is characterized by partial loss of myelin, axons, and oligodendroglial cells; mild reactive astrocytic gliosis; and sparsely distributed macrophages as well as stenosis resulting from hyaline fibrosis of arterioles and smaller vessels. These changes are

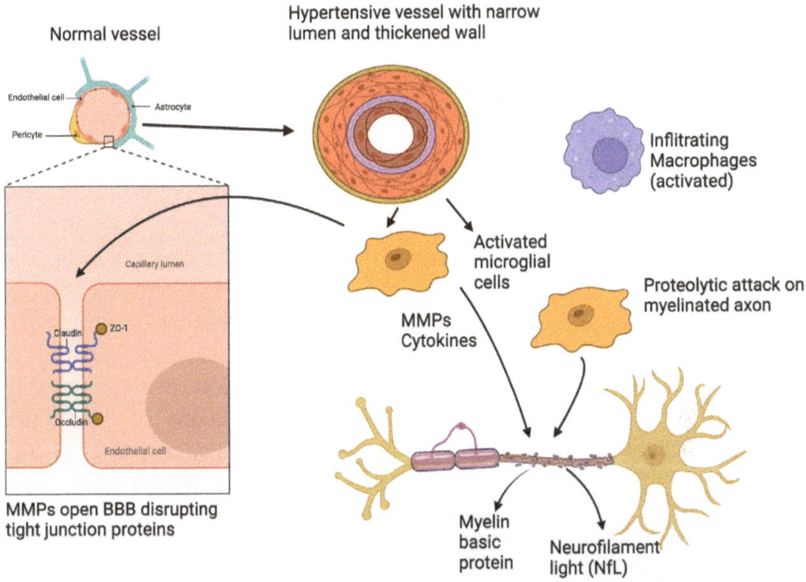

FIGURE 8.4 By-stander demyelination and BBB disruption by MMPs. Schematic illustration of by-stander demyelination mechanism in BD. A normal vessel is shown in the upper left. There are tight junction proteins between the endothelial cells that include claudin, occludin and zona occluden (zo-1). The vessel in the top center shows the effects of long-standing hypertension with narrowing of the lumen and thickening of the outer wall. Hypertensive vessels attract microglia/macrophages that attempt to remodel the vessel. These activated inflammatory cells secrete inflammatory factors, namely, MMPs and cytokines. These activated cells destroy the myelinated axons resulting in the release of myelin basic protein and neurofilament light (NfL), which can be measured in CSF and blood as a biomarker of inflammatory injury. Infiltrating macrophages (top right) can add to the proteolytic damage. Figure drawn with Bioender.

different than those found in complete or cavitating infarctions and hypertensive vascular changes are not observed. The white matter changes are thought to be due to hypoperfusion of the white matter territories since, in addition to the white matter hyaline vascular stenosis, these cases show signs of cardiovascular disease, usually with hypotension. The white matter disorder also occurs independent of the GM process of AD and may be seen as the sole brain lesion in non-AD subjects. Its occurrence is thus neither regularly related to the severity nor to the regional appearance and accentuation of the cortical Alzheimer process and is thus not likely to be just the result of a Wallerian degeneration. Histologically, it is similar in several respects to Binswanger's disease, although with some distinct differences. It is thus related to the cerebrovascular group of disorders. In view of its frequency and severity, this white matter lesion is important to define, to diagnose, and ultimately to prevent or cure (Brun & Englund, 1986). This type of pathological change may explain the finding of DTI abnormalities in the normal-appearing white matter since these pathological changes in white matter seen in AD are similar to those seen in vascular disease; they may have similar pathophysiology with hypoxic hypoperfusion initiating the incomplete infarction.

Microglia as source of persistent inflammation in Binswanger's disease

Microglia are the resident macrophages of the brain, and they are critical in innate immunity. Normally microglia are quiescent. They move around in the brain, surveying for signs of inflammation. Once they detect inflammation, they change shape into the rounded form of toxic macrophages that are capable of secreting a wide variety of substances that can damage the cell. After an acute injury, they are recruited to remove debris of dead cells. Patients with white matter disease have a progressive illness with gradually

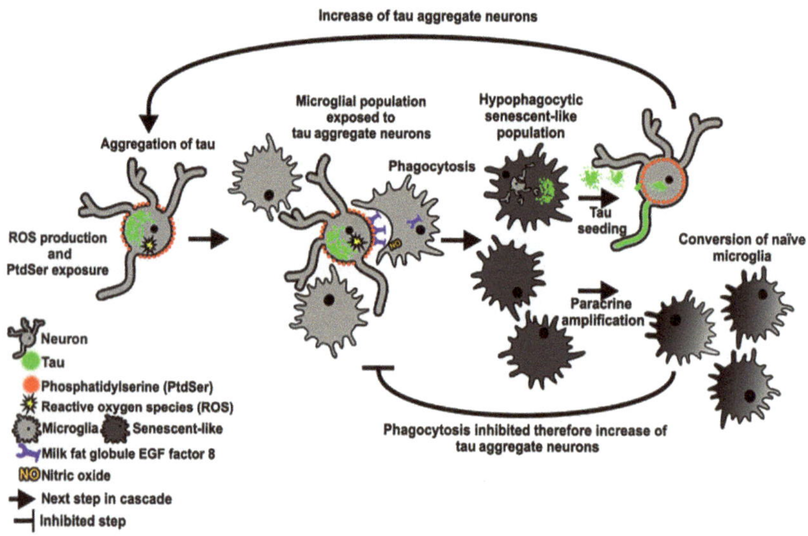

FIGURE 8.5 Schematic of proposed microglial dysfunction associated with phagocytosis of tau aggregate–containing neurons. Tau aggregate–containing neurons increase reactive oxygen species (ROS) production and expose PtdSer, which recruits cocultured microglia to perform phagocytosis. When phagocytosis occurs, the community of microglia undergoes phenotypic transitions characterized by hypophagocytic behavior and the acquisition of senescence-associated traits. Microglia that have phagocytosed tau aggregate–containing neurons secrete seeding-competent tau aggregates that can induce further tau aggregation, while paracrine factors can convert naïve microglia to become hypophagocytic/senescent-like. The end result is an increasing number of neurons with tau aggregates in a self-amplifying cycle. *Adapted from Brelstaff, J. H., Mason, M., Katsinelos, T., McEwan, W. A., Ghetti, B., Tolkovsky, A. M., & Spillantini, M. G. (2021). Microglia become hypofunctional and release metalloproteases and tau seeds when phagocytosing live neurons with P301S tau aggregates.* Science Advances, 7 *(43), eabg4980. https://doi.org/10.1126/sciadv.abg4980.*

increasing lesions in the white matter. These patients have evidence of inflammation in the CSF and brain, including disruption of the BBB. There is a diffuse pattern of injury that includes the regions with white matter hyperintensities. However, normal-appearing white matter, which are areas in which the white matter hyperintensities seen in FLAIR MRI are absent, also can have increased BBB permeability. This suggests that the inflammation is more diffuse with the ongoing release of toxic factors that open the barrier and attack the myelin. One mechanism to explain this chronic progressive damage to the white matter is the formation of senescent microglia that continue to secrete MMP-3. The mechanism for this to occur involves microglia that phagocytize injured neurons expressing pTau; these neurons express the injury marker, phosphatidylserine, which signals microglia to ingest the damaged neuron. The ingested neurons contain aggregates of tau that they spread as tau seeds in a prion-like spreading pattern. After phagocytosing tau aggregate containing neurons, the microglia enter a senescence phenotype, secreting senescence-associated cytokines along with matrix remodeling enzymes, particularly the active form of MMP-3. This could explain the role of senescent microglia in the diffuse leakage of the blood vessels and the ongoing damage to the myelinated fibers (Fig. 8.5) (Brelstaff et al., 2021).

References

Alvarez-Vergara, M. I., Rosales-Nieves, A. E., March-Diaz, R., Rodriguez-Perinan, G., Lara-Ureña, N., Ortega-de San Luis, C., Sanchez-Garcia, M. A., Martin-Bornez, M., Gómez-Gálvez, P., Vicente-Munuera, P., Fernandez-Gomez, B., Marchena, M. A., Bullones-Bolanos, A. S., Davila, J. C., Gonzalez-Martinez, R., Trillo-Contreras, J. L., Sanchez-Hidalgo, A. C., Del Toro, R., Scholl, F. G., … Pascual, A. (2021). Non-productive angiogenesis disassembles Aß plaque-associated blood vessels. *Nature Communications*, *12*(1), 3098. Available from https://doi.org/10.1038/s41467-021-23337-z.

Brelstaff, J. H., Mason, M., Katsinelos, T., McEwan, W. A., Ghetti, B., Tolkovsky, A. M., & Spillantini, M. G. (2021). Microglia become hypofunctional and release metalloproteases and tau seeds when phagocytosing live neurons with P301S tau aggregates. *Science Advances*, *7*(43). Available from https://doi.org/10.1126/sciadv.abg4980, eabg4980.

Brosnan, C. F., Cammer, W., Norton, W. T., & Bloom, B. R. (1980). Proteinase inhibitors suppress the development of experimental allergic encephalomyelitis. *Nature*, *285*(5762), 235–237.

Brun, A., & Englund, E. (1986). A white matter disorder in dementia of the Alzheimer type: A pathoanatomical study. *Annals of Neurology*, *19*(3), 253–262.

Chandler, S., Coates, R., Gearing, A., Lury, J., Wells, G., & Bone, E. (1995). Matrix metalloproteinases degrade myelin basic protein. *Neuroscience Letters*, *201*(3), 223–226.

Desai, B. S., Schneider, J. A., Li, J.-L., Carvey, P. M., & Hendey, B. (2009). Evidence of angiogenic vessels in Alzheimer's disease. *Journal of Neural Transmission (Vienna, Austria : 1996)*, *116*(5), 587–597. Available from https://doi.org/10.1007/s00702-009-0226-9.

Eeg-Olofsson, O., Link, H., & Wigertz, A. (1981). Concentrations of CSF proteins as a measure of blood brain barrier function and synthesis of IgG within the CNS in "normal" subjects from the age of 6 months to 30 years. *Acta Paediatrica Scandinavica*, *70*(2), 167–170.

Englund, E., Brun, A., & Alling, C. (1988). White matter changes in dementia of Alzheimer's type. Biochemical and neuropathological correlates. *Brain: A Journal of Neurology*, *111*(Pt 6), 1425–1439.

Fern, R., & Matute, C. (2019). Glutamate receptors and white matter stroke. *Neuroscience Letters*, *694*, 86–92. Available from https://doi.org/10.1016/j.neulet.2018.11.031.

Folkman, J. (2002). Role of angiogenesis in tumor growth and metastasis. *Seminars in Oncology*, *29*(6 Suppl 16), 15–18.

Förstermann, U., & Sessa, W. C. (2012). Nitric oxide synthases: Regulation and function. *European Heart Journal*, *33*(7), 829–837. Available from https://doi.org/10.1093/eurheartj/ehr304, 837a–837d.

Glenner, G. G. (1981). Amyloidosis. Its role in Alzheimer's disease and other diseases. *Annales de Pathologie*, *1*(2), 105–108.

Hansson, O., Santillo, A. F., Meeter, L. H., Nilsson, K., Landqvist Waldö, M., Nilsson, C., Blennow, K., van Swieten, J. C., & Janelidze, S. (2019). CSF placental growth factor – A novel candidate biomarker of frontotemporal dementia. *Annals of Clinical and Translational Neurology*, *6*(5), 863–872. Available from https://doi.org/10.1002/acn3.763.

Huisa, B. N., Caprihan, A., Thompson, J., Prestopnik, J., Qualls, C. R., & Rosenberg, G. A. (2015). Long-term blood-brain barrier permeability changes in Binswanger disease. *Stroke: A Journal of Cerebral Circulation*, *46*(9), 2413–2418. Available from https://doi.org/10.1161/STROKEAHA.115.009589.

Janelidze, S., Hertze, J., Nägga, K., Nilsson, K., Nilsson, C., Wennström, M., van Westen, D., Blennow, K., Zetterberg, H., & Hansson, O. (2017). Increased blood-brain barrier permeability is associated with dementia and diabetes but not amyloid pathology or APOE genotype. *Neurobiology of Aging*, *51*, 104–112. Available from https://doi.org/10.1016/j.neurobiolaging.2016.11.017.

Lo, E. H., Wang, X., & Cuzner, M. L. (2002). Extracellular proteolysis in brain injury and inflammation: Role for plasminogen activators and matrix metalloproteinases. *Journal of Neuroscience Research*, *69*(1), 1–9.

Nitkunan, A., Charlton, R. A., McIntyre, D. J. O., Barrick, T. R., Howe, F. A., & Markus, H. S. (2008). Diffusion tensor imaging and MR spectroscopy in hypertension and presumed cerebral small vessel disease. *Magnetic Resonance in Medicine*, *59*(3), 528–534. Available from https://doi.org/10.1002/mrm.21461.

Norton, W. T., Brosnan, C. F., Cammer, W., & Goldmuntz, E. A. (1990). Mechanisms and suppression of inflammatory demyelination. *Acta Neurobiologiae Experimentalis*, *50*(4–5), 225–235.

Norton, W. T., Cammer, W., Bloom, B. R., & Gordon, S. (1978). Neutral proteinases secreted by macrophages degrade basic protein: A possible mechanism of inflammatory demyelination. *Advances in Experimental Medicine and Biology*, *100*, 365–381.

Rosenberg, G. A. (2017). Extracellular matrix inflammation in vascular cognitive impairment and dementia. *Clinical Science (London, England: 1979)*, *131*(6), 425–437. Available from https://doi.org/10.1042/CS20160604.

Rosenberg, G. A., Sullivan, N., & Esiri, M. M. (2001). White matter damage is associated with matrix metalloproteinases in vascular dementia. *Stroke: A Journal of Cerebral Circulation*, *32*(5), 1162–1168.

Stadelmann, C., Timmler, S., Barrantes-Freer, A., & Simons, M. (2019). Myelin in the central nervous system: Structure, function, and pathology. *Physiological Reviews*, *99*(3), 1381–1431. Available from https://doi.org/10.1152/physrev.00031.2018.

Tarkowski, E., Issa, R., Sjögren, M., Wallin, A., Blennow, K., Tarkowski, A., & Kumar, P. (n.d.). Increased intrathecal levels of the angiogenic factors VEGF and TGF-beta in Alzheimer's disease and vascular dementia. *Neurobiology of Aging*, *23*(2), 237–243.

Tosto, G., Zimmerman, M. E., Carmichael, O. T., & Brickman, A. M. (2014). Predicting aggressive decline in mild cognitive impairment: The importance of white matter hyperintensities. *JAMA Neurology*, *71*(7), 872–877. Available from https://doi.org/10.1001/jamaneurol.2014.667.

Yang, Y., Kimura-Ohba, S., Thompson, J. F., Salayandia, V. M., Cossé, M., Raz, L., Jalal, F. Y., & Rosenberg, G. A. (2018). Vascular tight junction disruption and angiogenesis in spontaneously hypertensive rat with neuroinflammatory white matter injury. *Neurobiology of Disease*, *114*, 95–110. Available from https://doi.org/10.1016/j.nbd.2018.02.012.

Chapter 9

Machine learning and precision medicine

Introduction

Clinical medicine is experiencing a massive increase in the amount of information available to the physician caring for a patient. Certain medical fields have incorporated this deluge of information into patient care while others are lagging behind. Neurology has been slow to adopt the new methods to use the large amount of information, but it is rapidly learning from other fields. Cancer diagnosis and treatment has been at the forefront of this revolution; not only have there been an extensive number of genes associated with different cancers discovered, but this information has been used to formulate treatment plans. Other fields such as radiology and dermatology are using computer-aided imaging to diagnose illness by analysis of radiographs and to automate diagnoses of skin cancers. The concepts behind the use of machine learning in diagnosis originated from early work in the field of "cybernetics," which is a transdisciplinary approach for exploring regulatory systems—their structures, constraints, and possibilities. Norbert Wiener defined cybernetics in 1948 as "the scientific study of control and communication in the animal and the machine" (Wiener, 1948). From the early work on control theory by Wiener and others has slowly evolved the modern concepts of artificial intelligence (AI) and machine learning. There are various definitions of AI or machine learning. The term is used to describe computers that perform cognitive functions that we associate with the human mind; these "thinking machines" can beat experts in chess and the Chinese game of Go. In medicine, there are capable of analyzing large amounts of data to arrive at a diagnosis through pattern recognition. Antibiotic drugs have been designed by AI in ways that were unavailable to humans, pointing to the future of molecular discovery in medicine.

Machine learning

Machine learning is a fundamental concept of AI research since the field's inception is the study of computer algorithms that improve automatically through experience. Unsupervised learning is the ability to find patterns in a stream of input, without requiring a human to label the inputs first. Supervised

Neuroinflammation in Vascular Dementia. DOI: https://doi.org/10.1016/B978-0-12-823455-6.00005-5

learning includes both classification and numerical regression, which requires a human to label the input data first. Classification is used to determine what category something belongs in, and occurs after a program sees a number of examples of features from several categories, and learns to associate those features with a classification. Regression is the attempt to produce a function that describes the relationship between inputs and outputs and predicts how the outputs should change as the inputs change. Both classifiers and regression learners can be viewed as "function approximators" trying to learn an unknown (possibly implicit) function. Basic building blocks used in AI algorithms have been developed to learn from data. As they learn new strategies or heuristics they improve their function and can even write other algorithms. There are several basic approaches to building a learning system. These include Bayesian networks, decision trees, and nearest-neighbor. Much of AI research involves figuring out how to identify and narrow down a classifier to avoid considering a broad range of possibilities unlikely to be beneficial.

Approach of clinical medicine to diagnosis

Medical students are taught to make diagnoses based on a constellation of symptoms and signs. Traditionally, this involves two ways of thinking, which can be simply stated as Boolean or Bayesian. A Boolean approach combines a group of symptoms, compares them to a given pattern and gives the symptoms a name. For example, the symptoms headache, fever, and stiff neck would trigger a diagnosis of meningitis. This approach does not take into account the probability of a disease being present in a certain time and place. In New York, fever would not lead to suspicion of dengue fever or malaria, but would be more likely to be caused by influenza, particularly in the winter. The Bayesian approach requires information on the probability of a diagnosis given other circumstances; this models the way a clinician would approach the problem. Neither of these approaches is sufficient for diagnosis of dementia due to the similarity of the clinical presentations for many of the causes; additional diagnostic aids are needed. These aids can be as simple as blood tests to rule out medical conditions or as complicated as genetic testing. As the amount of information that can be used in the diagnostic process grows, the ability to make effective use of the information is overwhelming. This has lead to a number of approaches from making lists of possible symptoms associated with a specific diagnosis to use of AI, particularly when the number of features is very large.

Early attempts at improving diagnostic accuracy in dementia

One of the first studies at improving diagnosis in vascular dementia was a study of cerebral blood flow (Table 9.1) (Hachinski et al., 1975).

TABLE 9.1 Hacinski ischemic score.

Clinical feature	Score
Abrupt onset of dementia	2
Stepwise deterioration	1
Somatic complaints	1
Emotional incontinence	1
Hypertension (past or present)	1
History of stroke	2
Focal neurological symptoms	2
Focal neurological signs	2
(There are no intermediate scores for 2-point features i.e. either 0 or 2)	
A score of 2 or less are typical of a patient with Alzheimer's disease	
A score of greater than 2 is typical of multi-infarct dementia	
Recent research has shown that the Hachinski score has no predictive value when using autopsy as the gold-standard diagnostic tool	

The Hachinski ischemic score was used for many years as a means to separate the vascular dementia patients from those with Alzheimer's disease. However, it was of limited use since it was neither confirmed by autopsy studies nor useful beyond diagnosing vascular contributions. This is an example of a Boolean approach; however, it is possible to use the Bayesian probability methods with more accuracy. This score was constructed prior to the availability of computed tomography (CT) or magnetic resonance imaging (MRI) and without the aid of biomarkers obtained from cerebrospinal fluid (CSF) and positron emission tomography (PET). It represents a type of Boolean approach because a number of features trigger a diagnosis. Separating vascular features from Alzheimer's disease was possible with limited accuracy and it is now rarely used in research studies. It has been replaced with other methods of separating patients into diagnostic categories. As the newer methods of neuroimaging and CSF analysis became available in most centers, biomarkers became more prevalent.

Both large and small vessel disease (SVD) cause vascular cognitive impairment and dementia (VCID). This heterogeneity has lead to disagreements over the terms used to describe vascular dementia, impeding clinical trials. SVD is the main cause of white matter (WM) damage and a major factor in VCID (another term is subcortical ischemic vascular disease—SIVD). This is generally a progressive process, while large vessel disease, due to either major vessel thrombosis or emboli, occurs in an unpredictable pattern, making the natural history sporadic and leaving SIVD, which is progressive rather than random, the optimal form for treatment trials. SIVD of Binswanger's disease (BD) type is characterized by extensive involvement of the WM with neuropsychological abnormalities primarily in executive function and abnormalities in balance and continence. Differentiation of BD from other forms of neurodegeneration is often difficult and disease markers have been used to facilitate early diagnosis; they should appear in the early stages and be validated by long-term follow-up. There are a number of potential biomarkers to aid diagnosis of SIVD of the BD type, including WM hyperintensities (WMHs) on MRI, metabolite levels with proton magnetic resonance R spectroscopic imaging (^{1}H-MRSI), quantification of blood–brain barrier (BBB) permeability, executive dysfunction on neuropsychological testing, and elevated albumin ratio and matrix metalloproteinases (MMPs) in the CSF. However, used individually none of the biomarkers can reliably diagnose SIVD. A more productive approach is to combine clinical, imaging, and CSF studies in a multimodal process to improve diagnostic accuracy, particularly at an early stage when treatments are most likely to be beneficial. Such an approach was tested by obtaining biomarkers at entry into the study and comparing them with diagnoses made after multiyear follow-up (Rosenberg et al., 2015). This multimodal approach requires the use of computer analysis of combinations of clinical, imaging and CSF studies as disease markers of patients with VCID. The biomarkers selected were suggestive of neuroinflammation. The goal was to identify specific patients with clinical symptoms possibly related to neuroinflammation with BBB disruption and WM changes. Multiple factors were combined since none of the clinical, biochemical, or imaging features alone were sufficient. A hybrid approach was used with a 10-point score that was formed from clinical, multimodal imaging, and CSF results, resembling Boolean logic; each patient's Binswanger's disease score (BDS) was calculated based from a list similar to the Hachinski Score, which is based on the number of components in the scale.

Another independent approach to calculate a score for BD, was to use a statistical method called principal component analysis (PCA) and exploratory factor analysis (EFA) with nine of the most dominant factors that are most likely to be important biomarkers. In this early use of the statistical approach, the suspected dominant factors, including average T-score for executive tests, levels of three ^{1}H-MRS metabolites (NAA, CHO and CR), albumin ratio, MMP-2, MMP-9 (calculated by indexing the CSF levels to those in the blood), Aβ142/logP-Tau$_{181}$, and mean BBB permeability from DCE-MRI. The PCA/EFA analysis identified four factors that can be considered a distinct scale on its own. The EFA uses a correlation matrix of the variables, and to perform optimally, requires variables with an approximately normal distribution, which is best achieved by log transforming variables that are skewed. An initial PCA indicated that four variables could account

TABLE 9.2 Expanded features to identify patients with Binswanger's disease.

Features used in scale scores	BDS	PCA/EFA
I. Clinical characteristics (1 point each)		
1. HTN	X	
2. DM	X	
3. Hyper-reflexia (REF)	X	
4. Imbalance (GAIT)	X	
II. Neuropsychological testing (1 point)		
	X	X
III. Metabolites in WM (1H-MRSI) (1 point)		
6. NAA < 12[12]	X	X
7. Choline (CHO)		X
8. Creatine and phosphocreatine (CR)		X
IV. Inflammation and BBB (3 points)		
9. Albumin index >6.0 (albumin ratio)	X	X
10. BBB permeability Ki >0.0018[13]	X	X
11. MMP-2 index <0.01[18]	X	X
12. MMP-9 index		X
V. Alzheimer's biomarkers (1 point)		
13. Aβ42/log(P-τ181) >150	X	X

BBB, Blood–brain barrier; BDS, Binswanger's disease score; *DM*, diabetes mellitus; *EFA*, exploratory factor analysis; *HTN*, hypertension; *MMP*, matrix metalloproteinase.; *NAA*, *N*-acetylaspartate; PCA, principal component analysis; *WM*, white matter.

for the major variance in the data (Table 9.2). Using four variables to avoid missing possible patterns, we identified from the EFA the "loading" patterns defining the new factors (Erhardt et al., 2021).

Two models were tested for correlation of disease markers with final diagnosis. In model I, 10 disease markers were selected that were obtained at study entry for inclusion in the BDS, and in model II, PCA and EFA with quantitative biomarkers were used to calculate a probability score for having a diagnosis of BD. In the BDS model, points were given for each clinical feature present, and cut-points were used to assign points for continuous variables. To determine the correspondence of the final diagnoses with the BDS obtained at entry, we divided patients into two groups: less than six and greater than or equal to six. There were no between-group differences for age, diabetes or Aβ42/logP-Tau_{181}. However, hypertension, gait abnormalities, albumin ratio, MMP-2 index and mean permeability were statistically greater in the group with higher BDSs. Executive function T-scale scores were significantly lower. The second statistical approach more closely approximated the subsequent clinical diagnosis. This approach showed that PCA/EFA was an improvement. However, this again was limited to identifying those dementia patients with a vascular component, leaving out other types of dementia.

An important step forward occurred with the increasing use of neuroimaging, beginning in the 1980s, related to the increasing clinical use of MRI, which showed a high incidence of changes in the WM. This forced a reevaluation of the concept of multi-infarct dementia since the majority of patients lacked MRI evidence of discreet strokes, showing instead microvascular disease or small strokes. This resulted in a new concept that diffuse injury to the WM in itself could lead to cognitive impairment (Bowler, 2007). With the increased interest in understanding the diseases of the WM, the realization that many normal people had such changes and the concept of leukoaraiosis or WM changes of normal aging emerged (Hachinski et al., 1987). Newer MRI methods were used to show the injury of the WM due to ischemia: proton NMR spectroscopy was the earliest method used (Gasparovic et al., 2013).

NMR spectroscopy as a biomarker

Using the MRI machine as it was originally invented as a spectroscopic analysis instrument overcame the limitations of FLAIR MRI. ^{1}H-MRSI could be used to show individual molecules as a spectroscopic image. Protons were ideal for this purpose because they were ubiquitous on all molecules and in sufficient abundance to be detected by the relatively low sensitivity of NMR. The major molecules in the brain detected by ^{1}H-MRSI are *N*-acetylaspartate (NAA) and creatinine (Cr). Originally NAA was thought to be found exclusively in neurons, making it an excellent injury marker. However, the recent discovery of a higher concentration of NAA in myelin and oligodendrocytes than in neurons raises questions about the validity of the use of NAA as a neuronal marker of injury (Nordengen et al., 2015). Creatinine is important in brain energy. This recent finding adds to the potential for NAA to be used as a marker for WM injury. Reduced levels of NAA and Cr in the brains of patients with dementia correlates with neuropsychological test results, while the size of WM lesions on FLAIR MRI does not. However, NAA and Cr correlate with size of WMHs. While WMHs correlate with brain metabolites, neuropsychological test results only correlate with NAA and Cr, not WMHs (Gasparovic et al., 2013). Using advanced imaging methods, such as ^{1}H-MRSI, to calculate the level of NAA in the WM provides an improved biomarker compared to the use of WMHs on FLAIR. It is possible to create a table with the major features used in the diagnosis of VCID that can be analyzed by either counting the number of features as is done in the Hachinski ischemic score or using regression methods to find the principal components.

Diffusion tensor imaging

Diffusion tensor imaging (DTI) is an excellent method to demonstrate pathological changes in the WM most likely due to inflammation and ischemia. Compared to ^{1}H-MRSI, DTI is more accessible for routine clinical use and is the preferred method to show regions of injury; there is a high correlation between NAA signal and DTI for WM, but NAA requires an NMR physicist to run the sequences and interpret the results, which is not necessary for DTI. There are several ways to analyze DTI data: mean diffusivity (MD), radial diffusivity (RD), free water (FW), and peak width of skeletonized mean diffusivity (PSMD). They are data collected from the similar MRI sequences; the optimal one to use depends on multiple factors. Adding DTI data to the diagnostic pathways improves the ability to separate the various types of dementia; the damage to the WM as shown by DTI is a strong indicator that the patient has a form of vascular disease. Before it is possible to say that with certainty, it is necessary to rule out other causes of WM injury, particularly multiple sclerosis and other neuroinflammatory conditions. Testing the CSF for factors associated with multiple sclerosis and for infection is necessary to rule out non-ischemic etiologies. The BDS can be expanded to include the NAA and Cr data from the ^{1}H-Mrs. This larger number of factors can be used to calculate the Principle Component Analysis (PCA) and to compare it to the BDS. This approach is useful in identifying those patients most likely to have the ischemic form of dementia, but is too limited to use to separate the vascular from the neurodegenerative forms of dementia, and of little use when there is dual pathology of the mixed dementia patients. The factors involved in this type of analysis are shown in Table 9.2 (Rosenberg et al., 2015). The values for DTI can be substituted for ^{1}H-MRS. While this is an operational approach that has biological validity, it is incapable of handling the large increase in factors that have become available.

Computer advances and big data

Rapid advances in computer technology software and imaging hardware have had a major impact on multiple medical fields, including cancer diagnosis using genes, radiological imaging, and dermatological diagnoses in skin cancers. While the impact on diagnosis of neurological illnesses is at an early stage, there is much promise for rapid advances in computer miniaturization and improvement in algorithms. Imaging technology has made similar advances propelled by developments in the hardware used in video gaming programs that has made computer image recognition now a widely available technology. Computational software based on statistical packages that can analyze "big data," such as the widely used R software, are able to handle vast amounts of data efficiently, making it possible to incorporate many more features into diagnostic programs.

The ability of machine learning to advance diagnosis in neurological diseases, and equally important advances in methods to measure small amounts of abnormal proteins in blood and CSF, added to the progress. Methods for the analysis of body fluids based on ultra-sensitive assay systems have opened up another avenue of information. Measurements of proteins, RNA, and genes are now being done routinely with a wide array of methods; the major technique for protein analysis is the enzyme-linked immunosorbent assay (ELISA), a commonly used analytical biochemical method. First described by Engvall and Perlmann (1971). The assay uses a solid-phase type of enzyme immunoassay to detect the presence of a ligand in a liquid sample using antibodies directed against the protein to be measured. Multiplex immunoassays address workflow and sample volume problems by combining assays for many analyte targets in a single reaction volume, reducing workflow and sample volume problems. Multiplex platforms can offer distinct advantages over single plex assays. The ability of multiplex assays to use smaller analyte volumes, typically 25–50 μL, has made these the optimal assays. Additionally, multiplex platforms often provide a gain in assay dynamic range. Whereas, ELISA rapidly loses linearity over a few orders of magnitude, multiplex assays are reported to maintain linearity over three or even five orders of magnitude. As panels often contain targets that are present at radically different concentrations, linearity over a wide range is essential for reliable data. A number of methods are typically used for multiplex immunoassays, and they loosely fit into two categories: immobilization on a solid surface in which the assays for each analyte are spatially separated; immobilization on beads or particles in which the assays for each analyte are on a different bead/particle. Multiplex ELISAs can analyze over one hundred analytes simultaneously; a number of these assay kits are available commercially, allowing rapid measurements of proteins in CSF and blood with a high degree of accuracy. This can provide an overwhelming amount of data beyond the analysis capacity of routine statistical methods; these large datasets are ideal for the newer computer-based methods to analyze.

The problem is compounded when additional data from clinical evaluation, neuropsychological evaluation, and imaging are included, creating datasets that contain hundreds of pieces of information. Video gaming boards advanced the field of pattern recognition and AI by vastly expanding the ability to manipulate massive amounts of information. This ability solves the earlier problem of selecting the information from the various modalities to include in a scale where over 10 features became difficult to analyze. There is no limit to the amount of information that can be analyzed by programs, such as R, which are becoming routine in statistical analysis.

Combining different data modalities is essential if diagnoses that have overlapping features are to be used. For example, it is common to find several pathologies at autopsy, particularly in those with advanced age. Clinical diagnosis can make a best guess that a patient has multiple etiologies based on MRI findings and clinical testing. Biological biomarkers improve the accuracy of the clinical assessment. The key factors are to use the MRI to indicate vascular injury with FLAIR or DTI to document WM ischemia, and to use CSF or blood to show the presence of proteins related to neurodegeneration. In Alzheimer's disease research, the Alzheimer proteins are now required for the biological diagnosis; these can come from CSF or PET measurements of beta-amyloid or phosphorylated tau. Multietiology dementia is most commonly a combination of Alzheimer's disease and vascular disease. By adding MRI measures to the Alzheimer's disease diagnosis, a diagnosis of mixed dementia can be made during life. With this approach it is possible to diagnose pure Alzheimer's disease and pure vascular cognitive impairment, and to add the new diagnostic category of mixed dementia. Since the MRI indicates injury to the WM and the CSF or PET indicates neurodegenerative disease of the Alzheimer's type, they form the x and y axes that can be used with a double dichotomy approach to separate patients, depending on location in a feature space (Caprihan et al., 2021). An example of the method used is shown in Fig. 9.1. The double dichotomy method to classify patients with dementia is correlated with executive function (Fig. 9.2).

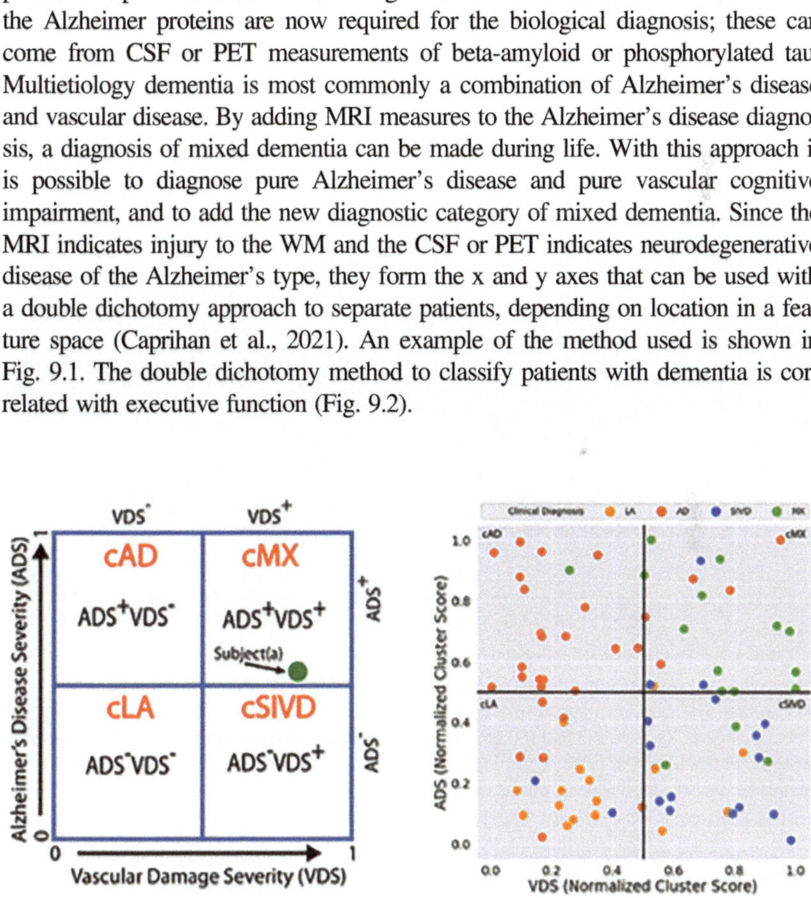

FIGURE 9.1 (Left) The double-dichotomy clustering method is a two-dimensional scatter plot with the *x*-axis being the VDS score and the y-axis being the ADS score, and a cut-off of 0.5 for each axis gives four quadrants, with each quadrant describing subjects with different disease characteristics. The four cluster (c)-based patient groups are cLA = VDS$^-$ ADS$^-$, cAD = VDS$^-$ ADS$^+$, cSIVD = VDS$^+$ ADS$^-$, and cMX = VDS$^+$ ADS$^+$. The location of a subject in this plot describes disease severity. For example, subject (a) is classified as cMX, but being close to the ADS boundary, we know that its ADS factors are similar to some of the SIVD subjects (Right). *ADS*, Alzheimer's disease score; *SIVD*, subcortical ischemic vascular disease; *VDS*, vascular disease score. *Drawing by Arvind Caprihan. Permission given to use in book.*

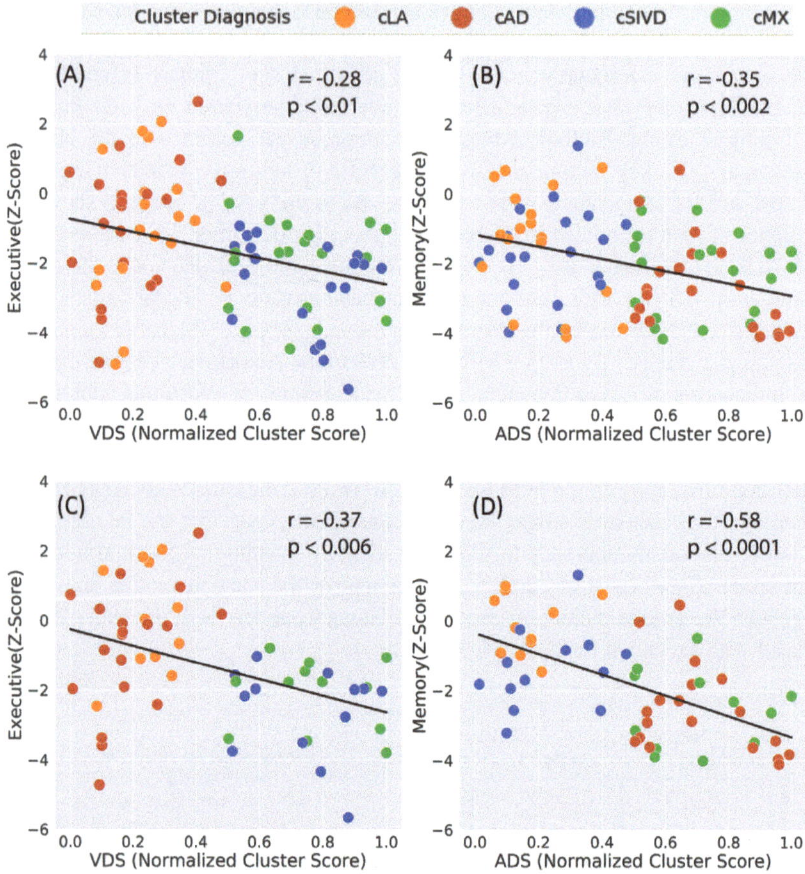

FIGURE 9.2 The spearman correlation between the executive function and the VDS score and between the memory function and the ADS score for the full patient group (80 subjects) is shown in parts (A) and (B), and for a subset of the subjects (55 subjects) in parts (C) and (D). The selected subset of subjects for parts (C) and (D) are those with matching clinical and biomarker-based cluster diagnosis. In this subset of subjects there is a stronger correlation with cognitive function. This is an example of how cluster diagnosis can be used to identify a more homogenous group of subjects with well-characterized properties. *From Caprihan, A. unpublished data with permission.*

The spearman correlation between the executive function and the vascular disease score (VDS) and between the memory function and the Alzheimer's disease score (ADS) for the full patient group (80 subjects) is shown in Fig. 9.2A and B and for a subset of the subjects (55 subjects) in Fig. 9.2C and D. The selected subset of subjects for Fig. 9.2C and D are those with matching clinical and biomarker-based cluster diagnosis. In this subset of subjects there is stronger correlation with cognitive function. This is an example of how cluster diagnosis can be used to identify a more homogenous group of subjects with well characterized properties.

The Alzheimer Disease Neuroimaging Initiative (ADNI) is a collective database containing data from all the Alzheimer's disease research centers (ADRCs). The data in the ADNI database contains both MRI and CSF data, making it possible to calculate the VDS and ADS and to perform a double dichotomy calculation to separate patients by biological criteria rather than clinical data. The main advantage of the ADNI database is the large number of subjects that can be included in the analysis. The disadvantage is the data was collected for analysis of Alzheimer's disease, and patients with vascular disease that would fall into the SIVD category were excluded. Application of the double dichotomy analysis to the ADNI database separates patients into the major dementia categories, and allows for the determination of mixed dementia during life (Fig. 9.3). The double dichotomy approach provides biological diagnoses that use biomarkers from MRI diffusion tenor imaging and CSF Alzheimer's disease proteins. Currently, the ADS is obtained with the invasive lumbar puncture. However, it may be possible to use plasma with the ultra-sensitive assay systems being developed to bypass the CSF

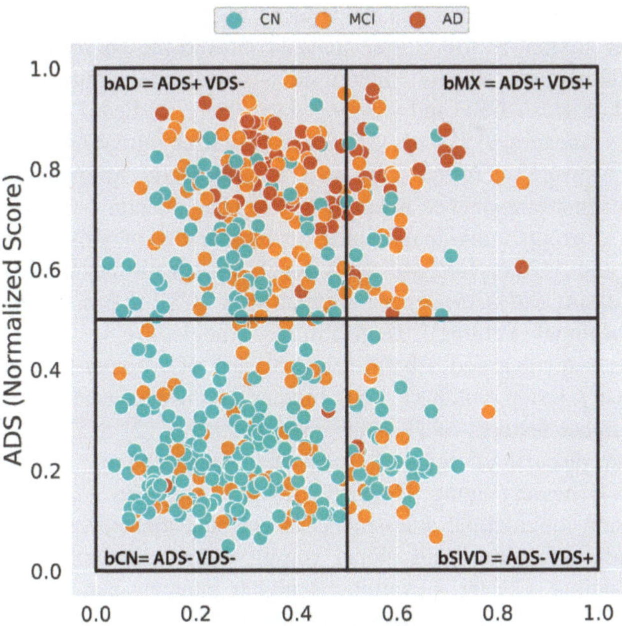

FIGURE 9.3 Expanded double dichotomy based on the ADNI database. Participants were CN, MC, or AD. The upper left corner is bAD. The upper right quadrant is bMX. The lower quadrants are bCN and bSIVD. *AD*, Alzheimer's disease; *ADNI*, Alzheimer Disease Neuroimaging Initiative; *bAD*, biological Alzheimer's disease; *bCN*, biological cognitive normal; *bMX*, biological mixed dementia; *bSIVD*, biological subcortical ischemic vascular disease; *CN*, clinically normal; *MC*, mild cognitive impairment.

and use MRI and plasma, which would make the use of the biological diagnostic approach available to a larger population (Fig. 9.4).

Big data in neurological diagnosis

Cluster analysis

There are several approaches that can be used with large datasets: the data can be used to form groups *a priori* through a clustering method or the patients can be grouped by clinical criteria with the features determined in the groups. Clustering methods have the advantage that they allow the features that have been measured to determine which features are close to each other. In one study, the investigators identified a subgroup of patients with AD characterized by very high levels of CSF total-tau and phosphorylated-tau and low levels of Aβ42. This subgroup of patients exhibited a more malignant disease course, including faster progression of cognitive deficits, worse outcome of cholinesterase treatment, and a higher mortality rate (Wallin et al., 2010). This approach used two biomarkers to show the most involved patients with the greatest risk of progressing had elevated pTau and reduced Aβ.

Another attempt to use cluster analysis showed the ability to detect preclinical Alzheimer's disease: they used a multimodal approach including neuropsychological, CSF and imaging data. They studied 175 middle-aged participants (mean age 55.9 $\pm$ 5.7 years at first cognitive assessment, 70% female) that were recruited from two longitudinally followed cohorts to undergo magnetic resonance imaging and lumbar puncture. Cluster analysis was used to group individuals based on biomarkers of amyloid pathology (CSF amyloid-β42/amyloid-β40 assay levels), magnetic resonance imaging-derived measures of neurodegeneration/atrophy (CSF-to-brain volume ratio, and hippocampal volume), neurofibrillary tangles (CSF phosphorylated $Ptau_{181}$ assay levels), and a brain-based marker of vascular risk (total WM hyperintensity lesion volume). Four biomarker clusters emerged consistent with preclinical features of (1) Alzheimer's disease; (2) mixed Alzheimer's disease and vascular etiology; (3) suspected non-Alzheimer's disease etiology; and (4) healthy aging. Cognitive decline was then analyzed between clusters using longitudinal assessments of episodic memory, semantic memory, executive function, and global cognitive function with linear mixed effects modeling. Cluster 1 exhibited a higher intercept and greater rates of decline on tests of episodic memory. Cluster 2 had a lower intercept on a test of semantic memory and both Cluster 2 and Cluster 3 had steeper rates of decline on a test of global cognition. Additional analyses on Cluster 3, which had the smallest hippocampal volume, suggest that its biomarker profile is more likely due to hippocampal vulnerability and not to detectable specific volume loss exceeding the rate of normal aging. These results demonstrate that pathology, as indicated by biomarkers, in a preclinical timeframe is related to patterns of longitudinal cognitive decline. Such biomarker patterns may be useful for identifying at-risk populations to recruit for clinical trials. This study provides empirical support for the use of cluster analysis to detect heterogeneity in biomarker profiles and potentially identify a subgroup with preclinical Alzheimer's disease in late midlife. Racine et al. identified four biomarker groups and then investigated clinical interpretations of the groups by examining differential cognitive decline: they identified a group that appears to represent healthy aging as well as a group whose biomarker and cognitive profiles are consistent with preclinical Alzheimer's disease. Specifically, the preclinical Alzheimer's disease cluster presented with elevated CSF phosphorylated Tau_{181}, low CSF amyloid-β42/amyloid-β40, and greater decline on tests of episodic memory. Consistent with an Alzheimer's-like profile, this cluster was also older and had a greater incidence of *APOE4*. Clusters did not differ in other demographic or clinical variables examined (Racine et al., 2016).

Agglomerative hierarchical clustering analysis (HCA) is a commonly used unsupervised machine learning approach for identifying informative natural clusters of observations. HCA is done by calculating a pairwise dissimilarity matrix and then clustering similar observations until all observations are grouped within a cluster. Verifying the empirical clusters produced by HCA is complex and not well studied in biomedical applications. In one study, the authors demonstrated the comparability of a novel HCA technique with one that was used in previous biomedical applications while applying both techniques to plasma angiogenic (FGF, FLT, PIGF, Tie-2, VEGF, VEGF-D) and inflammatory (MMP1, MMP3, MMP9, IL8, TNFα) protein data to identify informative subsets of individuals. Study subjects were diagnosed with mild cognitive impairment due to cerebrovascular disease (MCI-CVD) (Fig. 9.4). Through comparison of the two HCA techniques, they were able to identify subsets of individuals, based on differences in VEGF ($P < 0.001$), MMP-1 ($P < 0.001$), and IL-8 ($P < 0.001$) levels.

These profiles provide novel insights into angiogenic and inflammatory pathologies that may contribute to VCID. The usage of both the novel HCA model and a Euclidean distance HCA model identified a novel subset of patients within the MCI-CVD population. This study provides insight into a potential underlying inflammatory and angiogenic profile of disease in patients with VCID. Defining subsets of patients within this population with different disease profiles continues to be a key research objective. These profiles can provide a more complete understanding of disease progression and allow physicians and researchers to identify patients undergoing different rates of pathologic change in a prospective cohort. Further studies are needed to clarify these profiles by combining plasma and MRI imaging biomarkers that can also be used in clinical trials as key outcome measures to determine the efficacy of novel therapeutics (Fig. 9.5) (Winder et al., 2020).

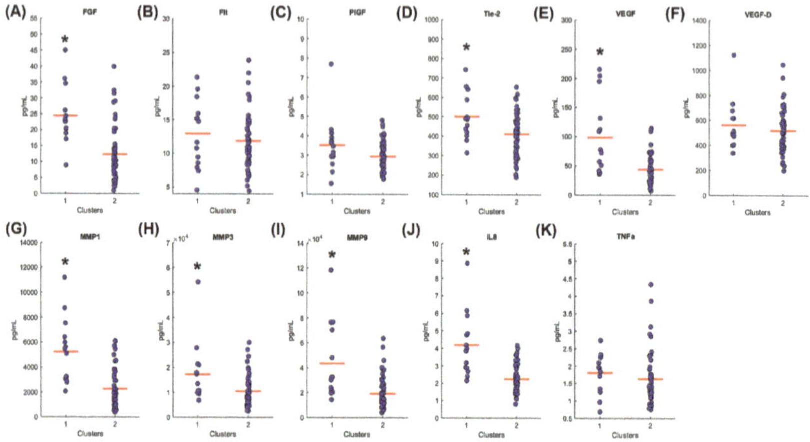

FIGURE 9.4 **Individual biomarkers for cluster analysis in VCID.** Compares clusters 1–4 produced from the Euclidean distance model in each inflammatory and angiogenic protein measured. (A–K) Show mean and scatter of FGF (A), FLt (B), PIGF (C), Tie-2 (D), VEGF (E), VEGF-D (F), MMP-1 (G), MMP-3 (H), MMP-9 (I), IL-8 (J), and TNF-α (K). Red horizontal lines indicate the means for each cluster. Stars (*) indicate statistical significance between groups as calculated by the log transform of the data shown using ANOVA followed by Tukey's HSD for post hoc analysis (P < 0.05). *VCID*, Vascular cognitive impairment and dementia. *From Winder, Z., Sudduth, T. L., Fardo, D., Cheng, Q., Goldstein, L. B., Nelson, P. T., Schmitt, F. A., Jicha, G. A. & Wilcock, D. M. (2020). Hierarchical clustering analyses of plasma proteins in subjects with cardiovascular risk factors identify informative subsets based on differential levels of angiogenic and inflammatory biomarkers.* Frontiers in Neuroscience, 14, *84. https://doi.org/10.3389/fnins.2020.00084*

Proteomics

A number of new instruments have become available that allow for simultaneous measurements of proteins. These proteomic assays can measure over 100 analytes in tiny samples of CSF and plasma. Proteomics complements the instruments used to measure genes, providing data that can be used to construct pathways of multiple genes. These large datasets produce several challenges in the selection of the optimal statistical methods to use and in the interpretation of the findings. Generally, multiple linear regression methods are used initially and complemented with machine learning programs based on the R statistical package to identify the features of maximal importance in the statistical results. When 10 or more proteins of importance are identified, the more complicated task is to assign a meaningful scenario for their action, particularly when they fall into different metabolic pathways. Another major challenge is to identify the groups that will be included in the final analysis; clustering sidesteps this complication by using the data to form the groups, but when multiple diagnostic categories are involved and the desire is to identify the features that fall within each of the groups, the

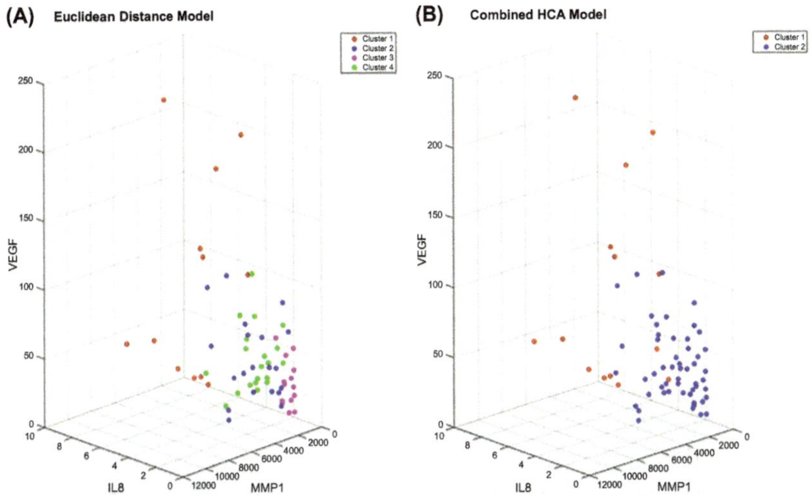

FIGURE 9.5 **Cluster comparison of inflammatory biomarkers in VCID.** Cluster comparison of VEGF (pg/mL), MMP1 (pg/mL), and IL8 (pg/mL) for data clustered using the Euclidean distance model (A) and the combined HCA model (B) from 66 participant plasma samples. *HCA*, Hierarchical clustering analysis; *VCID*, vascular cognitive impairment and dementia. *From Winder, Z., Sudduth, T. L., Fardo, D., Cheng, Q., Goldstein, L. B., Nelson, P. T., Schmitt, F. A., Jicha, G. A. & Wilcock, D. M. (2020). Hierarchical clustering analyses of plasma proteins in subjects with cardiovascular risk factors identify informative subsets based on differential levels of angiogenic and inflammatory biomarkers.* Frontiers in Neuroscience, 14, *84. https://doi.org/10.3389/fnins.2020.00084.*

problems encountered are compounded. Because of the limitations in constructing meaning from the results, the disease groups are often limited to one disease with multiple levels of severity and comparisons are made with a normal cohort.

An early paper using a proteomic approach focused on stages of Alzheimer's disease based on the Clinical Dementia Rating (CDR) scale, which goes from 0 in normals to 1 in MCI, the mildly affected. They used the Luminex platform to analyze 190 analytes. Comparing the results to Aβ and Tau proteins, they identified several biomarkers that enhanced the ability to separate the normals from the early AD. Using Luminex technology and a targeted multiplex panel, they identified 37 analytes (12 with Bonferroni correction) that are increased or decreased in the CSF of participants with early AD relative to cognitively normal controls. ROC analysis revealed that small combinations of a subset of these markers (cystatin C, VEGF, TRAIL-R3, PAI-1, PP, NT-proBNP, MMP-10, MIF, GRO-α, fibrinogen, FAS, and eotaxin-3) can enhance the ability of the best-performing of the traditional biomarkers, the tau/Aβ42 ratio, to discriminate CDR 0.5 and 1 from CDR 0 participants (Craig-Schapiro et al., 2011).

Another multiplex proteomics study of CSF and plasma used the Olink ProSeek immunoassay to measure 270 CSF and plasma proteins in a large cohort of patients. They identified, and replicated, altered levels of ten CSF proteins in Aβ + individuals, including CHIT1, SMOC2, MMP-10, LDLR, CD200, EIF4EBP1, ALCAM, RGMB, tPA and STAMBP. They also identified and replicated alterations of six plasma proteins in Aβ + individuals OSM, MMP-9, HAGH, CD200, AXIN1, and uPA. Multiple analytes associated with cognitive performance and cortical thickness ($q < 0.05$). Plasma biomarkers could distinguish AD dementia and prodromal AD from CN. This study showed the importance of contributions of immune markers, phospholipids, angiogenic proteins and other biomarkers downstream of, and potentially orthogonal to, Aβ and tau in AD, and identified candidate biomarkers for earlier detection of neurodegeneration (Whelan et al., 2019).

Most of the studies of large numbers of proteins in CSF and plasma have focused on different stages of AD; they move beyond the standard biomarkers of Aβ and tau to include a variety of inflammatory and angiogenic factors. While they are inclusive of proteins from different classes of biomarkers, they have yet to provide a coherent picture of the interaction of these proteins which is essential in reconstructing a pathophysiology that could lead to therapeutic interventions. This is likely to occur as the use of higher levels of computer analysis is employed, particularly the involvement of AI for identifying the patterns of interaction that are important in the various pathways.

References

Bowler, J. V. (2007). Modern concept of vascular cognitive impairment. *British Medical Bulletin*, *83*(1), 291–305. Available from https://doi.org/10.1093/bmb/ldm021.

Caprihan, A., Raja, R., Hillmer, L. J., Erhardt, E. B., Prestopnik, J., Thompson, J., Adair, J. C., Knoefel, J. E., & Rosenberg, G. A. (2021). A double-dichotomy clustering of dual pathology dementia patients. *Cerebral Circulation - Cognition and Behavior*, 2. Available from https://doi.org/10.1016/j.cccb.2021.100011.

Craig-Schapiro, R., Kuhn, M., Xiong, C., Pickering, E. H., Liu, J., Misko, T. P., Perrin, R. J., Bales, K. R., Soares, H., Fagan, A. M., & Holtzman, D. M. (2011). Multiplexed immunoassay panel identifies novel CSF biomarkers for Alzheimer's disease diagnosis and prognosis. *PLoS One*, *6*(4), e18850. Available from https://doi.org/10.1371/journal.pone.0018850.

Engvall, E., & Perlmann, P. (1971). Enzyme-linked immunosorbent assay (ELISA). Quantitative assay of immunoglobulin G. *Immunochemistry*, *8*(9), 871–874.

Erhardt, E. B., Adair, J. C., Knoefel, J. E., Caprihan, A., Prestopnik, J., Thompson, J., Hobson, S., Siegel, D., & Rosenberg, G. A. (2021). Inflammatory biomarkers aid in diagnosis of dementia. *Frontiers in Aging Neuroscience*, *13*717344. Available from https://doi.org/10.3389/fnagi.2021.717344.

Gasparovic, C., Prestopnik, J., Thompson, J., Taheri, S., Huisa, B., Schrader, R., Adair, J. C., & Rosenberg, G. A. (2013). 1H-MR spectroscopy metabolite levels correlate with executive function in vascular cognitive impairment. *Journal of Neurology, Neurosurgery and Psychiatry*, *84*(7), 715–721. Available from https://doi.org/10.1136/jnnp-2012-303878.

Hachinski, V. C., Iliff, L. D., Zilhka, E., Du Boulay, G. H., McAllister, V. L., Marshall, J., Russell, R. W., & Symon, L. (1975). Cerebral blood flow in dementia. *Archives of Neurology*, *32*(9), 632–637.

Hachinski, V. C., Potter, P., & Merskey, H. (1987). Leuko-araiosis. *Archives of Neurology*, *44*(1), 21–23. Available from https://doi.org/10.1001/archneur.1987.00520130013009.

Nordengen, K., Heuser, C., Rinholm, J. E., Matalon, R., & Gundersen, V. (2015). Localisation of *N*-acetylaspartate in oligodendrocytes/myelin. *Brain Structure & Function*, *220*(2), 899–917. Available from https://doi.org/10.1007/s00429-013-0691-7.

Racine, A. M., Koscik, R. L., Berman, S. E., Nicholas, C. R., Clark, L. R., Okonkwo, O. C., Rowley, H. A., Asthana, S., Bendlin, B. B., Blennow, K., Zetterberg, H., Gleason, C. E., Carlsson, C. M., & Johnson, S. C. (2016). Biomarker clusters are differentially associated with longitudinal cognitive decline in late midlife. *Brain: A Journal of Neurology*, *139*(Pt 8), 2261–2274. Available from https://doi.org/10.1093/brain/aww142.

Rosenberg, G. A., Prestopnik, J., Adair, J. C., Huisa, B. N., Knoefel, J., Caprihan, A., Gasparovic, C., Thompson, J., Erhardt, E. B., & Schrader, R. (2015). Validation of biomarkers in subcortical ischaemic vascular disease of the Binswanger type: Approach to targeted treatment trials. *Journal of Neurology, Neurosurgery and Psychiatry*, *86*(12), 1324–1330. Available from https://doi.org/10.1136/jnnp-2014-309421.

Wallin, A. K., Blennow, K., Zetterberg, H., Londos, E., Minthon, L., & Hansson, O. (2010). CSF biomarkers predict a more malignant outcome in Alzheimer disease. *Neurology*, *74*(19), 1531–1537. Available from https://doi.org/10.1212/WNL.0b013e3181dd4dd8.

Whelan, C. D., Mattsson, N., Nagle, M. W., Vijayaraghavan, S., Hyde, C., Janelidze, S., Stomrud, E., Lee, J., Fitz, L., Samad, T. A., Ramaswamy, G., Margolin, R. A., Malarstig, A., & Hansson, O. (2019). Multiplex proteomics identifies novel CSF and plasma biomarkers of early Alzheimer's disease. *Acta Neuropathologica Communications*, *7*(1), 169. Available from https://doi.org/10.1186/s40478-019-0795-2.

Wiener, N. (1948). Cybernetics. *Scientific American*, *179*(5), 14–18.

Winder, Z., Sudduth, T. L., Fardo, D., Cheng, Q., Goldstein, L. B., Nelson, P. T., Schmitt, F. A., Jicha, G. A., & Wilcock, D. M. (2020). Hierarchical clustering analyses of plasma proteins in subjects with cardiovascular risk factors identify informative subsets based on differential levels of angiogenic and inflammatory biomarkers. *Frontiers in Neuroscience*, *14*, 84. Available from https://doi.org/10.3389/fnins.2020.00084.

Chapter 10

Future directions

Medicine has been a major benefactor of the advances in computer science and data analysis. Biomarkers are now available to enhance diagnosis in many diseases with multiple assays on single plates for analysis with ultrasensitive instruments. The genetic background of many diseases can be routinely determined by whole genome sequencing to identify patterns of gene expression unique to each disease. Neurology uses these advanced methods in many diseases. The genetic basis of many neurological diseases is routinely used to aid diagnosis, and the use of highly sensitive chemical analysis methods has identified a number of biomarkers characteristic of different types of dementia. Large datasets obtained from neuroimaging, plasma and cerebrospinal fluid (CSF) protein analysis, provide insights into disease mechanisms that will guide development of new therapies. It is still unclear how the availability of all this information will impact day to day clinical practice because of the cumbersome analyses requiring computers to only be available to solve complex diagnostic problems over longer timeframes. Much will depend on the speed of the analysis machines and the ability to enter data into computers for the application of machine learning. Clearly, the path to precision medicine is visible, and the means to achieve it are already here. They are altering our thinking about the pathophysiology of dementia hold great promise in giving direction to future therapies.

Amyloid reconsidered

The dominance of the amyloid theory over research in dementia has been shaken by large clinical trials that removed amyloid from the brain without affecting cognition. This has prompted a search for alternate theories for Alzheimer's pathology; while aging has the major impact, investigators are studying the role of vascular disease and inflammation related to multiple misfolded proteins that accumulate in the brain. Aging increases inflammation and has a negative effect on blood vessels, which develop narrowing of the lumen and fibrosis of the outer walls, reducing cerebral blood flow and vessel responsiveness. Microglia are involved in the inflammation by secreting proteases and cytokines. Pericytes, the cells embedded in the basal lamina of blood vessels, function as microglia/macrophages. Microglia can be detected

Neuroinflammation in Vascular Dementia. DOI: https://doi.org/10.1016/B978-0-12-823455-6.00006-7

with novel ligands for use in positron emission tomography (PET) scans. Magnetic resonance imaging (MRI)/PET scanners can measure blood–brain barrier (BBB) permeability with a gadolinium contrast agent, and at the same time observe the presence of microglia. The combined MRI BBB and PET microglia study showed that they were not active in the same regions, suggesting that they may be separate processes, which leaves unanswered the question of the role of microglia in vascular damage. Another study of patients with Alzheimer's disease (AD) using multiple PET ligands along with CSF triggering receptor expressed on myeloid cells 2 (*TREM2*) gene, a marker of microglia inflammation and brain tissue, found that the microglia ligand, [^{11}C] PBR28, correlated with CSF soluble TREM2 and showed regional distribution resembling *TREM2* gene expression in brain tissue. More importantly, the microglial activation and tau PET correlated hierarchically with each other following Braak-like stages. These findings support a model where an interaction between amyloid-β (Aβ) and activated microglia sets the pace for tau spread across Braak stages (Pascoal et al., 2021).

Misfolded proteins that are trapped in the brain initiate inflammation. There are a number of abnormal proteins in addition to amyloid present in the brains of cognitively impaired patients at autopsy: these include α-synuclein, phosphoTau, and transactive response DNA-binding protein 43 (TDP-43) (Karanth et al., 2020). These misfolded proteins interfere with normal function. Although these proteinopathies are identified with specific diseases, when they are found individually with increasing age there is a buildup of multiple pathologies. An autopsy study of patients at University of Kentucky that had been studied for up to 10 years prior to death provided estimates of the incidence of each of these types of pathology: the final sample included 375 individuals [mean age at death of 86.9 (8.0) years; mean (SD)]. Quadruple misfolded proteins were detected in 41 of 214 individuals with dementia (19.2%). Overall, 46 individuals (12.3%) had quadruple misfolded proteins, whereas 143 individuals (38.1%) had 3 proteinopathies. Dementia frequency was highest among those with quadruple misfolded proteins [41 (89.1%)], and participants with quadruple misfolded proteins had the lowest final mean (SD) mini-mental status exam (MMSE) scores of 13.4 (9.8) points. The *APOE* ε4 allele was associated with higher odds of quadruple misfolded proteins (Karanth et al., 2020).

Mean (SD) time spent in the mild cognitive impairment (MCI) state was shortest among those in the quadruple misfolded protein group [1.7 (0.6) years] and among those with initial normal cognition [1.8 (0.6) years]. This finding suggests a more aggressive disease course for individuals with quadruple misfolded proteins. Cerebrovascular burden (atherosclerosis, arteriolosclerosis, and infarcts) was similar in all groups with the highest numbers in those with atherosclerosis in the quadruple misfolded proteins group (Karanth et al., 2020). The surprising aspect of this study was that 50% of the brains had at least 3 proteinopathies, and all had in addition some level of cerebrovascular disease. This is consistent with the concept that a single pathological protein may not necessarily lead to cognitive impairment, but that multiple abnormal proteins, which most likely increase the inflammatory burden, are more damaging. Furthermore, those with a greater number of abnormal proteins had a more rapid clinical decline, indicating the significance of accumulated abnormal proteins. Vascular disease was present in most of the brains, supporting the importance of mixed dementia (MX).

Advanced magnetic resonance imaging methods

Resting-state functional magnetic resonance imaging

Functional MRI (fMRI) is an advance in MRI that has shown brain activity as a relationship to cerebral blood flow following various stimulation paradigms. In addition, there is a pattern of brain activity in the fMRI even when there is no stimulation. That is called resting-state functional MRI (rs-fMRI). Brain activity can be observed because cerebral blood vessels respond to levels of oxygen and carbon dioxide by a change in caliber that affects cerebral blood flow. Blood oxygen level–dependent (BOLD) contrast imaging is an imaging method that detects a change in the relative levels of oxyhemoglobin and deoxyhemoglobin, created by a differential magnetic susceptibility that can be detected by MR imaging. The change in the BOLD signal forms the basis for fMRI. Low frequency produced by neural activity fluctuations can be seen in the BOLD signal, making it possible to identify regions that are correlated in their activity. The first application of this was the demonstration of a highly correlated BOLD time course between the left somatosensory cortex and the homologous areas in the contralateral hemisphere during the performance of bilateral finger tapping (Biswal et al., 1995). Task-based fMRI has been used to identify cortical regions mainly in normal subjects but is of limited use in patients with cognitive impairment who cannot perform specific tasks associated with the paradigms. rs-fMRI is a potential alternative modality that does not require task performance. There are many ways to analyze rs-fMRI data (Fig. 10.1) (Lv et al., 2018).

Structurally, the brain is organized grossly into different regions specialized for processing and relaying neural signals; functionally, the brain is subspecialized for perceptual and cognitive processes. Structural and functional brain abnormalities have been widely identified in dementia, but with variable replicability and significant overlap. Alzheimer's disease (AD) and Binswanger's disease (BD) share similar clinical symptoms and common brain changes on structural MRI (sMRI) that can confound diagnosis. Some of these difficulties can be overcome by the use of rs-fMRI. Going one step beyond rs-fMRI and combining rs-fMRI with diffusion MRI (dMRI) provides further insights into the two diseases. One study aimed to investigate correlated structural and functional brain changes in AD and BD by

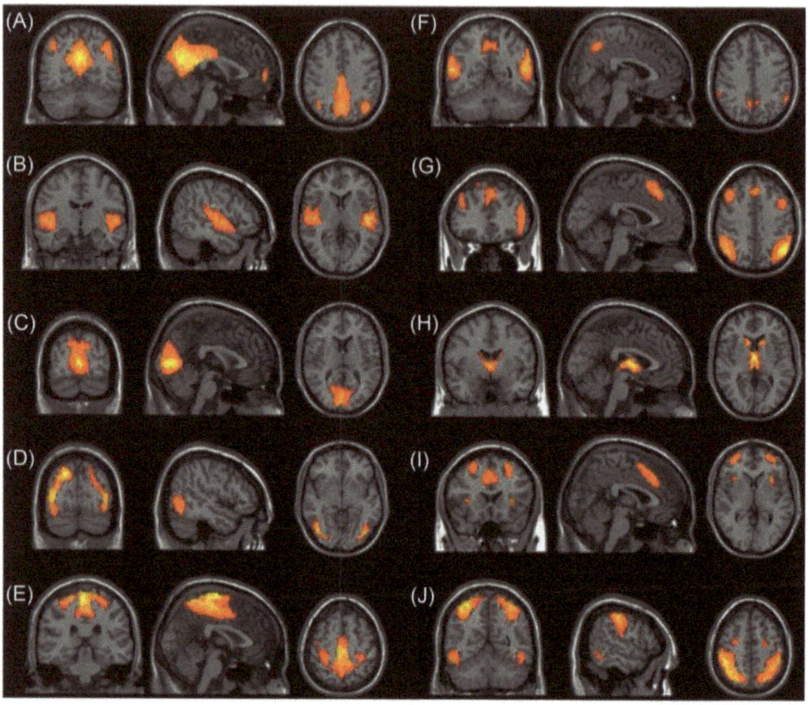

FIGURE 10.1 Typical reported networks (results of 200 healthy volunteers). (A) Default-mode network. (B) Auditory network. (C) Medial visual network. (D) Lateral visual network. (E) Sensorimotor network. (F) Precuneus network. (G) Dorsal visual stream (frontoparietal attention network). (H) Basal ganglia network. (I) Executive control network. (J) Visuospatial network. *From Lv, H., Wang, Z., Tong, E., Williams, L. M., Zaharchuk, G., Zeineh, M., Goldstein-Piekarski, A. N., Ball, T. M., Liao, C. & Wintermark, M. (2018). Resting-state functional MRI: Everything that nonexperts have always wanted to know.* AJNR. American Journal of Neuroradiology, 39*(8), 1390–1399. https://doi.org/10.3174/ajnr.A5527.*

combining rs-fMRI and dMRI. The method involved first performing a group independent component analysis on the fMRI data to extract 49 intrinsic connectivity networks (ICNs). Then a second analysis was conducted that included a multiset canonical correlation on three features: functional network connectivity (FNC) between ICNs, fractional anisotropy (FA) and mean diffusivity (MD). This revealed that two inter-correlated components showed significant group differences. The first component demonstrates distinct brain changes between AD and BD: AD shows increased cerebellar FNC, but decreased thalamic and hippocampal FNC. Such FNC alterations are linked to the decreased corpus callosum FA. AD also has increased MD in the frontal and temporal cortex, but BD shows opposite alterations. The second component demonstrates specific brain changes in BD. Increased FNC is mainly between default mode and sensory regions, while decreased FNC is mainly within the default-mode domain and related to auditory regions. The FNC changes are associated with FA changes in the posterior/middle cingulum cortex and visual cortex and increased MD in thalamus and hippocampus. They conclude that the findings provide evidence of linked functional and structural deficits in dementia and suggest that AD and BD have both common and distinct changes in white matter (WM) integrity and functional connectivity (Fu et al., 2019).

Adding structure to rs-MRI in Alzheimer's disease and Binswanger's disease

Additional neuroimaging features are needed to distinguish between AD and BD in order to address the challenge of separating these two diseases due to overlapping symptoms that occur with a high prevalence in both diseases in elderly individuals. MRI neuroimaging techniques, such as fMRI, sMRI and dMRI provide more specific and biologically based approaches to probe brain alterations in dementia. Although WM injury on sMRI is a prominent brain marker for BD, about 30% of typical AD patients show similar lesions. dMRI assesses WM ultrastructure by detecting abnormal diffusion in radiographically normal appearing regions, making it more sensitive to WM injury than T2 and fluid-attenuated inversion recovery (FLAIR) on sMRI.

AD is characterized by widespread gray matter changes and disconnection of cortical and subcortical regions. The default-mode network, which is one of the most widely studied resting-state networks, is significantly affected by AD with patients showing decreased resting-state activity in the default-mode and hippocampal regions, suggesting disrupted connectivity between these regions.

Newer methods for analysis of MRI-based methods combine multimodal features to improve information obtained from a single entity because of the ability to detect potentially hidden relationships between different imaging modalities. Data fusion is a multimodal alternative strategy which analyzes data jointly that can effectively detect multimodal brain changes related to diseases. Data fusion, which involves multiset canonical correlation analysis (MCCA), is one of the promising methods for the exploration of co-impairments in multimodal imaging data. Specifically, it is a data-driven method that decomposes features of each modality into spatial maps and their corresponding canonical variants, and aims to identify co-varying brain patterns among two or more modalities by maximizing the inter-modality correlations between canonical variants. The superior performance of MCCA in achieving jointly associated multimodal components have been shown in comprehensive simulation studies. Although MCCA is primarily designed to find associations in multimodalities, it is also suitable for identifying joint alterations among different modalities in diseases. Taken together, the MCCA method is believed to be a powerful tool for finding the potential linkages between multimodalities and assessing the brain impairments associated with these linked components. An example of the use of MCCA in dementia is the fMRI and dMRI data fusion in patients with AD and BD. By using these methods, the separation of three diagnostic groups was achieved: controls, AD, and BD (Fig. 10.2) (Fu et al., 2020).

The results showed that (1) brain changes in fMRI FNC, dMRI FA, and dMRI MD are highly correlated with each other; and (2) AD and BD share both similar and different (and in some cases opposite) functional and structural brain changes. The overall results demonstrate that investigating multimodal brain imaging data through fusion analysis might help to capture more relevant information about brain disorders and advance our knowledge of the relationships between functional and structural abnormalities in the brain.

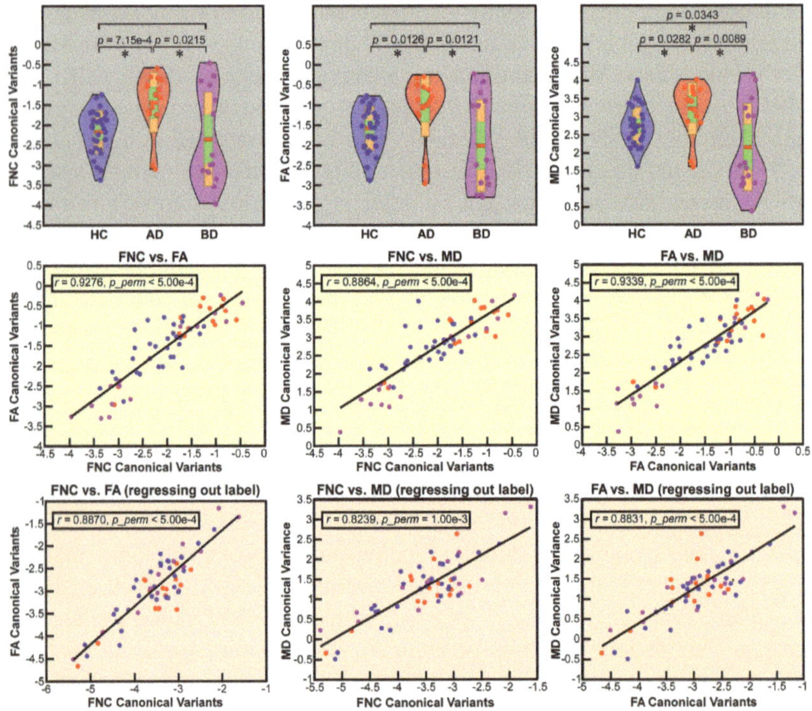

FIGURE 10.2 Group comparisons of canonical variants of joint component 1 and correlations between canonical variants. Top row: Boxplot of canonical variants of each feature component, with *p* values of generalized linear model (GLM) test between pair-groups (stars indicate significant group difference after False discovery rate [FDR] correction). Middle row: Pearson correlations between canonical variants of each feature component and *p* values of the permutation test. Bottom row: Pearson correlations between canonical variants of each feature component after regressing out group labels and *P* values of the permutation test. *FNC*, functional network connectivity; *FA*, fractional anisotropy; *MD*, mean diffusivity; *HC*, healthy controls; *AD*, Alzheimer's disease; *BD*, Binswanger's disease. *From Fu, Z., Iraji A., Caprihan, A., Adair, J. C., Sui, J., Rosenberg, G. A., & Calhoun, V. D. (2020). In search of multimodal brain alterations in Alzheimer's and Binswanger's disease.* NeuroImage Clinical, 26 *101937. Available from https://doi.org/10.1016/j.nicl.2019.101937. Epub 2019 Jul 15. PMID: 31351845; PMCID: PMC7229329.*

Genetic studies on early- and late-onset Alzheimer's disease

AD can be separated into early-onset and late-onset; both types have a genetic component. Most people with AD have the late-onset form of the disease, in which symptoms become apparent in their mid-60s and later. Researchers have not found a specific gene linked to late-onset AD. However, having a genetic variant of the apolipoprotein E (APOE) gene on chromosome 19 increases a person's risk. The *APOE* gene is involved in making a protein that helps carry cholesterol and other types of fat in the bloodstream. *APOE* comes in several different forms, or alleles. Each person inherits two *APOE* alleles, one from each biological parent. *APOE* ε2 is relatively rare and may provide some protection against the disease. If AD occurs in a person with this allele, it usually develops later in life than in someone with the *APOE* ε4 gene. *APOE* ε3, the most common allele, is believed to play a neutral role in the disease – neither decreasing or increasing risk. *APOE* ε4 increases risk for AD and is also associated with an earlier age of disease onset. About 25% of people carry one copy of *APOE* ε4; having one or two *APOE* ε4 alleles increases the risk of developing Alzheimer's. However, inheriting an *APOE* ε4 allele does not mean that a person will definitely develop Alzheimer's. Some people with an *APOE* ε4 allele never get the disease, and others who develop Alzheimer's do not have any *APOE* ε4 alleles.

ApoE has an important role in Aβ metabolism. *APOE* genotypes affect deposition of Aβ to form senile plaques. ApoE is co-deposited in senile plaques in the brains of AD patients, which is more abundant in *APOE* ε4 carriers compared with noncarriers. Between ages 50 and 59 years, 40.7% of *APOE* ε4 carriers had senile plaques compared with 8.2% of noncarriers (Kok et al., 2009). In individuals with positive Pittsburgh compound B (PiB)-PET images, which indicate fibrillar aggregates of Aβ, *APOE* ε4 was more common than in those with negative scans (65% vs 22%) in patients with AD (Barthel et al., 2011; Klunk et al., 2004). In a study of the effect of the *APOE* genotype on functional brain connectivity, it was found the gene was modulated by gender in healthy older human adults; they observed significantly decreased connectivity in the default-mode network in healthy older *APOE* ε4 carriers compared with ε3 homozygotes. There was an interaction between *APOE* genotype and gender: female ε4 carriers showed significantly reduced default-mode connectivity compared with either female ε3 homozygotes or male ε4 carriers, whereas male ε4 carriers differed minimally from male ε3 homozygotes. The higher prevalence of the ε4 allele among women with AD demonstrates that this interaction between *APOE* genotype and gender is detectable in the preclinical period. Presence of the *APOE* ε4 allele is also associated with increased risk for cerebral amyloid angiopathy and age-related cognitive decline during normal aging. ApoE–lipoproteins bind to several cell-surface receptors to deliver lipids and also to hydrophobic Aβ peptide, which is thought to initiate toxic events that lead to synaptic dysfunction and neurodegeneration in AD. *APOE* isoforms differentially regulate Aβ aggregation and clearance in the brain, and have distinct functions in regulating brain lipid transport, glucose metabolism, neuronal signaling, neuroinflammation, and mitochondrial function (Liu et al., 2013).

Some families carry genes that cause the early onset of a hereditary form of the illness. Early-onset AD is rare, representing less than 10% of all people with Alzheimer's. In those with the dominantly inherited form of the illness, the disease can occur between ages 30 and 60. Three genes are known to be present in early-onset AD: these are single-gene mutations associated with amyloid precursor protein (*APP*) on chromosome 21; presenilin 1 (*PSEN1*) on chromosome 14; and presenilin 2 (*PSEN2*) on chromosome 1. Mutations in these genes result in the production of the abnormal proteins associated with the disease. Each of these mutations plays a role in the breakdown of APP, a protein whose precise function is not yet fully understood. This breakdown is part of a process that generates toxic forms of amyloid plaque proteins.

AD biomarkers have been available for many years to aid in the diagnosis of AD; these include levels of amyloid and tau in the CSF and amyloid and tau PET. While these were accurate, CSF required an invasive lumbar puncture, and PET scanning was expensive and only done at research centers. The result was that most clinicians relied on clinical criteria that were confounded by the overlapping of clinical symptoms, which limited the precision of the diagnoses and interfered with clinical studies. More recently, biological criteria for use in clinical studies have been proposed. A biological diagnosis for use in research studies has been proposed; this approach uses AD proteins in the CSF or on PET scans along with brain atrophy. This rigorous definition is based on the formula ATN where A is amyloid, T is tau, and N is neurodegeneration (Jack et al., 2018). The ATN formula uses the major biomarkers in the CSF and PET along with MRI evidence of atrophy; AD diagnosis is based on the levels of proteins in the CSF, particularly the ratio of $A\beta_{1-42}$ to $A\beta_{1-40}$ and phosphoTau_{181}. Other factors could be added to the ATN formula to consider the discovery of additional biomarkers. Vascular disease is the most prevalent additional factor that could be added with the benefit of adding the MX category. It is likely that other biomarkers will be discovered in the future, and they could be incorporated into the formula.

Early diagnosis of Alzheimer's disease with measurement of Alzheimer's disease proteins in plasma

One of the factors attributed to the failure of clinical trials in AD is the late stage of the disease in which the drugs were tested. This has led to new strategies to identify the disease before cells have died and treatment would be fruitless. The development of ways to identify disease activity in the brain by the use of plasma assays has opened the potential for early diagnosis based on screening of large populations with the goal of finding those with abnormal proteins in the blood. A method to identify biomarkers in blood, which has been an aspirational goal, is now available from several companies. The development of ultrasensitive methods to measure very low levels of proteins has made the use of blood samples possible. Highly reliable automated measurements of amyloid and tau in the plasma are available with a single molecule array (SIMOA) that has femtogram per milliliter (fg/mL) level sensitivity for measurement of low abundance biomarkers in plasma. Unique technology provides robust detection for single analytes and selects multiple panel options. Other methods that can be used in plasma are also available as companies compete to take advantage of the large potential market. In addition to the AD proteins, a number of other platforms have been developed to measure a wide variety of proteins in the plasma. These are mainly multiplex ELISA arrays that are being used to measure proteases, angiogenic factors, and cytokines. One of these multiplex ELISA instruments is the MesoScale Discovery Electrochemiluminescence instrument which has the ability to detect multiple analytes in small sample volumes. These and other advances in measurement of plasma have made the detection of AD and other degenerative processes possible at earlier stages of the disease. Newer forms of tau proteins have improved the diagnostic accuracy of the plasma. In a cross-sectional study that included 1402 participants from 3 different AD patient cohorts, plasma P-tau_{217} discriminated AD from other neurodegenerative diseases in both neuropathologically and clinically defined cohorts with performance that was significantly better than established AD plasma- and MRI-based biomarkers but not significantly different from key CSF- or PET-based biomarkers. The discriminative accuracy of plasma P-tau_{217} for clinical AD dementia vs other neurodegenerative diseases was significantly higher than plasma P-tau_{181}, plasma neurofilament light (NfL), and MRI measures, but not significantly different compared with CSF P-tau_{217}, CSF P-tau_{181}, and tau PET. Furthermore, plasma P-tau_{217} levels were significantly greater among *PSEN1* mutation carriers, compared with noncarriers, from approximately 25 years and older, which is 20 years prior to estimated onset of MCI among mutation carriers (Fig. 10.3) (Thijssen et al., 2020).

Several of these plasma-based assays are already in use and others are being developed. Along with the improved methods for early diagnosis and staging of the various types of dementia illnesses, there are now available multiplex assay kits that can measure many proteins that are involved in the pathophysiology of the brain. While the multiplex assay kits provide information on many proteins that are suspected to be participating in dementia, the amount of information obtained generally requires a machine learning approach with artificial intelligence algorithms to interpret the data.

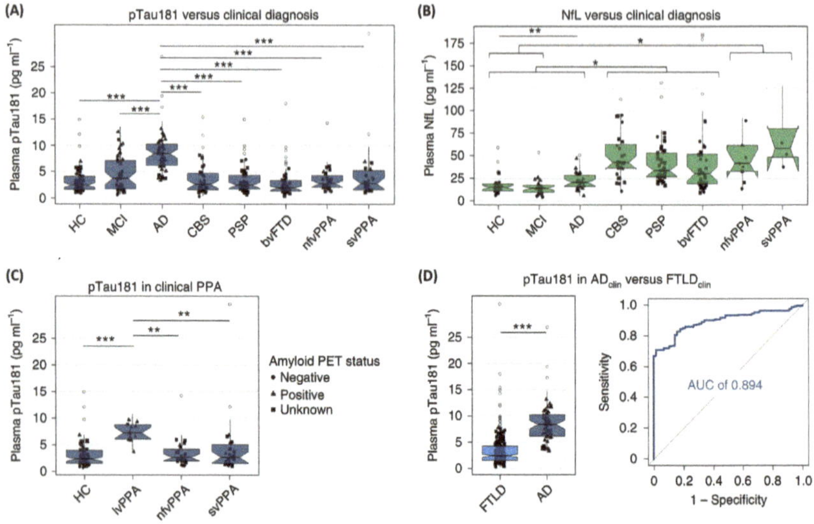

FIGURE 10.3 **Plasma pTau$_{181}$ and plasma NfL per clinical diagnosis.** (A) pTau$_{181}$ levels were elevated in ADclin compared to non-AD clinical diagnoses ($n = 362$). (B) Plasma NfL was lower in HCs and patients with MCI and AD compared to CBS, PSP and bvFTD, and NfL levels in HC and MCI were lower than in patients with nfvPPA and svPPA ($n = 213$). (C) Plasma pTau$_{181}$ levels are elevated in logopenic variant of primary progressive aphasia (lvPPA), which is typically caused by AD, as compared to levels in nonfluent/agrammatic variant of primary progressive aphasia (nfvPPA) and semantic variant of primary progressive aphasia (svPPA), which are typically caused by FTLD and HC ($n = 136$). (D) Plasma pTau$_{181}$ concentrations were increased in ADclin cases compared to FTLD clinical diagnoses and could differentiate between these groups ($n = 246$). The notch displays the 95% CI around the median. The shape reflects amyloid-PET status. ***$P < .0001$, **$P < .01$, *$P < .05$. *AD*, Alzheimer's disease; *bvFTD*, behavioral variant FTD; *CBS*, corticobasal syndrome; *CI*, confidence interval; *FTLD*, frontotemporal dementia; *HC*, healthy control; *MCI*, mild cognitive impairment; *NfL*, neurofilament light; *PET*, positron emission tomography; *PSP*, progressive supranuclear palsy. *From Thijssen, E. H., La Joie, R., Wolf, A., Strom, A., Wang, P., Iaccarino, L., Bourakova, V., Cobigo, Y., Heuer, H., Spina, S., VandeVrede, L., Chai, X., Proctor, N. K., Airey, D. C., Shcherbinin, S., Duggan Evans, C., Sims, J. R., Zetterberg, H., Blennow, K., … Boxer, A. L. (2020). Diagnostic value of plasma phosphorylated tau181 in Alzheimer's disease and frontotemporal lobar degeneration.* Nature Medicine, 26*(3), 387–397. https://doi.org/10.1038/s41591-020-0762-2.*

Cerebrospinal fluid and plasma neurofilament light as marker of axonal injury

Diseases of the WM are characterized by breakdown of the myelinated axons. Fragments of the myelinated fibers have been measured in the CSF to aid in the diagnosis of patients with multiple sclerosis. Myelin basic protein can be detected in the CSF and has been used to show that the axon is damaged. Another biomarker of axonal damage is the NfL protein, which is elevated in both CSF and plasma when the myelin is attacked (Khalil et al., 2018). Neurofilament light (NfL) is similar to myelin basic protein, and it can be used as a biomarker. The advantage of NfL is that it can be detected in the plasma bypassing CSF, which is needed for measurement of myelin basic protein. The main drawback of NfL is that it is elevated in a number of neurological conditions, making it a poor marker for specific disease processes. AD is one of the diseases in which NfL is increased; it is detected in both plasma and CSF by an ultrasensitive Quanterix assay. Both CSF and plasma NfL concentrations are a biomarker of damage to large-caliber myelinated axons. A study tested whether CSF NfL concentration is associated with cognitive decline and imaging evidence of neurodegeneration and WM change in AD; the study correlated baseline CSF NfL concentration with several longitudinal measures, including progression of cognitive impairment, WM change, and regional brain atrophy in several dementia diagnostic groups. Cerebrospinal fluid NfL concentration was elevated in AD, stable MCI, and progressive MCI group compared with control participants. In the MCI group, a higher CSF NfL concentration was associated with faster brain atrophy over time as measured by changes in whole-brain volume, ventricular volume, and hippocampus volume, faster disease progression as reflected by a decrease over time in Mini-Mental State Examination scores and increased Alzheimer Disease Assessment Scale cognitive subscale scores. Cerebrospinal fluid NfL concentration was increased early in AD and is associated with predicted cognitive deterioration and structural brain changes. This finding corroborates the contention that degeneration of large-caliber axons is an important feature of AD neurodegeneration (Zetterberg et al., 2016). In addition, WM injury as reflected on MRI correlates with CSF levels of NfL (Jonsson et al., 2010).

Hypoxia and inflammation augment the amyloid theory of chronic subcortical dementia

Innate immunity is important in inflammation in the brain; intrinsic microglia cells and recruited macrophages from systemic circulation both play a major role in perpetuating the inflammatory response. Activation of microglia takes place when products of tissue damage are present. Damage-associated molecular patterns (DAMP) are molecules within cells that are a component of the innate immune response released from dying cells damaged by trauma, ischemia or infection. Excessive amounts of abnormally folded proteins that stagnate within cells can trigger the pathological reaction. Both amyloid and tau are able to initiate cell death though this mechanism. Microglia respond to DAMPs through a series of Toll-like receptors on the cell surface. Signaling through the nuclear factor-κB (NF-κB) and AP-1 transcription factors, they initiate the formation of a complex collection of molecules that is called the inflammasome (Swanson et al., 2019). Central to this process is the NLPR3 molecule complex that aggregates along with the ASC spects. This complex molecular structure activates caspase-1 and begins activation of the cytokines,

IL-18 and IL-1β, both of which are toxic to cells, contributing to cell death. Strategies to block the formation and action of inflammasomes are emerging as possible therapies for many brain disorders. The confluence of multiple pathological processes, all of which have the potential to increase inflammation, creates a complex problem of determining the major ones that should be treated. The various proteinopathies have as an end result an increase in inflammatory biomarkers; compounding this plethora of proteins is the damage to the cerebral vasculature that initiates inflammation on its own. There are many sophisticated tools that have been developed for identifying the impact of inflammation, using MRI, blood and CSF.

Biomarkers associated with blood–brain barrier disruption

During injury and repair, while a number of molecular pathways are involved, three stand out as of major importance: angiogenic growth factors, proteases, and cytokines. In each of these major classes of molecules there are a number that have been shown to be important in dementia, but it has been difficult to determine for the individual molecules in each of the classes whether they have detrimental or beneficial effects. Molecules involved in angiogenesis include vascular endothelial growth factor-C (VEGF-C), vascular endothelial growth factor-D (VEGF-D) and their cognate receptors., Flt-1 and Tie-2. Placental growth factor has been detected in blood and CSF in dementia patients. Hypoxia increases the activity of the angiogenic growth factors to increase blood delivery, but the newly sprouting vessels are permeable, promoting edema. This dual function complicates understanding the beneficial and detrimental effects of their actions on both AD and vascular disease, since the angiogenic factors could be involved in both injury and repair of the brain.

Matrix metalloproteinases (MMP) are a family of 26 enzymes that are important in disruption of the extracellular matrix, which includes the basal lamina around blood vessels as well as the tight junction proteins that bind endothelial cells together. Multiplex ELISAs have become available that can multiple measure MMP-1, MMP-2, MMP-3, MMP-9 and MMP-10. Studies in patients with AD, subcortical ischemic vascular disease (SIVD), and mixed dementia (MX) have shown a pattern of MMP expression in chronic diseases that differs from that in acute ones. In the chronic situation, the CSF shows MMP-1, MMP-2, MMP-3 and MMP-10, but not MMP-9 (Erhardt et al., 2021). Further studies will be needed to determine what each of these MMPs is doing in a chronic situation.

Several of the biomarkers are associated with opening of the BBB, depending on the method used to measure permeability. Albumin index is associated with MMP-2, MMP-3, and MMP-10 in the proteases. The angiogenic factors are VEGF-C and Flt-1, while the cytokines are IL-6, IL-8 and tumor necrosis factor (TNF)-α, all of which are inflammatory cytokines.

Dynamic contrast-enhanced MRI (DCE-MRI), the gadolinium-based method to measure permeability is not correlated with the albumin index, indicating that the two methods are showing independent aspects of permeability. DCE-MRI is correlated with MRI biomarkers, including diffusion tensor imaging mean free water and PSMD along with WM hyperintensities volume. Albumin is 60 kDa molecular weight, while gadolinium is 40 Da. The larger molecule is most likely going through the blood vessel by pinocytosis, and the smaller molecule could go through an open tight junction. In both cases, the relationship of the biomarkers supports the biomarkers as indicators of inflammation.

Cytokines can have either inflammatory or anti-inflammatory actions. Much more is known about their role in inflammation, where they act in concert to form networks. One of these networks involving IL-18 has been linked with damage to the WM and cognitive dysfunction. Other molecules than cytokines are part of that network, including MPO, GDF-15, RAGE, ST2, and MCP-1. The STRING database query of the analytes reveals a biologically interconnected network centered on IL-18 and highly related to inflammation. Using an inflammatory composite score composed on a number of inflammatory biomarkers, the components of the inflammatory network can be discovered. This type of analysis is shown as colored nodes (bold) while first level interacting proteins are shown as white nodes (Fig. 10.4). Combining the inflammatory factors into a composite score, showed that they contributed to changes in the diffusion biomarker for free water and vascular risk factors (Fig. 10.5) (Altendahl et al., 2020). IL-18 and IL-1β are both proteolytic products formed from the action of inflammasomes.

Microglia, IL-1β, IL-6, and inflammasomes: drivers of white matter injury

WM disease is the hallmark of multiple sclerosis and small vessel disease and both are linked to inflammation. Despite the apparent similarities, they have different genetic backgrounds. A study of genetic databases for multiple sclerosis and small vessel disease found no overlapping genes, suggesting different links to inflammation. It could be that inflammatory mechanisms have no causal role in small vessel disease (Brown et al., 2019). Single nucleotide polymorphisms (SNPs) are associated significantly with immune pathways in MS, while SNPs that associate significantly with WMH (white matter hyperintensities) lesions principally code for extracellular matrix proteins and cell adhesion molecules (Brown et al., 2019). This suggests different immune pathways are relevant because of the differing nature of the primary insult rather than a shared mechanism of WM damage in ischemia and Ms; it may be that the inflammatory activation within a defined Ms lesion is mediated by a different pathway than the more generalized cellular response to hypoxia and differing metalloprotease expression between

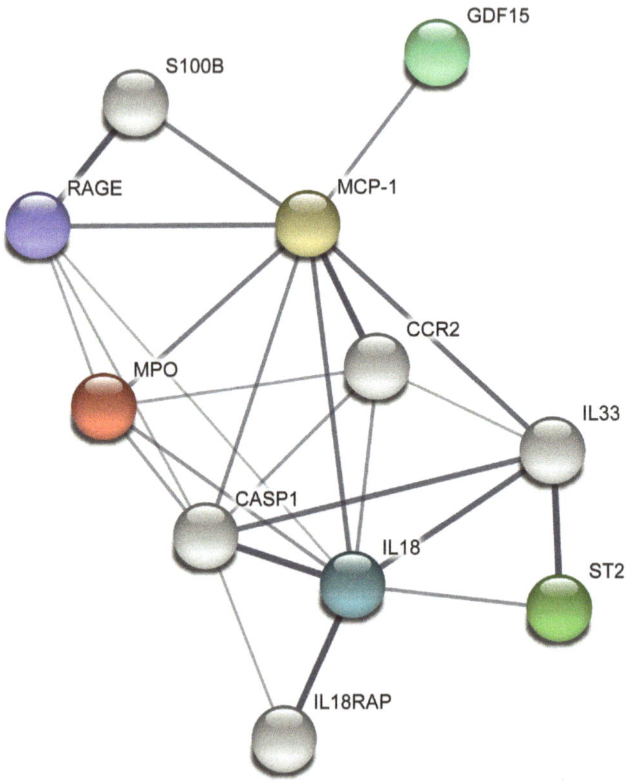

FIGURE 10.4 **ICS components form an inflammatory network.** The STRING database query of the six ICS component analytes reveals a biologically interconnected network centered on IL-18 and highly related to inflammation (*P*-value for protein interactions = 0.00022). ICS component analytes are shown as colored nodes (bold) while first level interacting proteins are shown as white nodes. Line width reflects the strength of data support. *From Altendahl, M., Maillard, P., Harvey, D., Cotter, D., Walters, S., Wolf, A., Singh, B., Kakarla, V., Azizkhanian, I., Sheth, S. A., Xiao, G., Fox, E., You, M., Leng, M., Elashoff, D., Kramer, J. H., Decarli, C., Elahi, F. & Hinman, J. D. (2020). An IL-18-centered inflammatory network as a biomarker for cerebral white matter injury.* PloS One, 15*(1), e0227835. https://doi.org/10.1371/journal.pone.0227835.*

ischemic and Ms lesions as has been shown (Anthony et al., 1997). In addition, two traits which do not share genetic risk might nevertheless act via the same cellular pathway activated by distinct triggers. It is also possible that common biological mechanisms might have opposite associations, leading to overall null results. Microglia are the main cells involved in innate immunity. Acting like systemic macrophages, microglia secrete the proteases and cytokines that lead to cell death. However, similar to all aspects of inflammation, the microglia exhibit both beneficial and detrimental aspects. There are ligands that can be used in PET scans to show activated microglia. PET/MRI scans can show both BBB disruption with Gadolinium using the

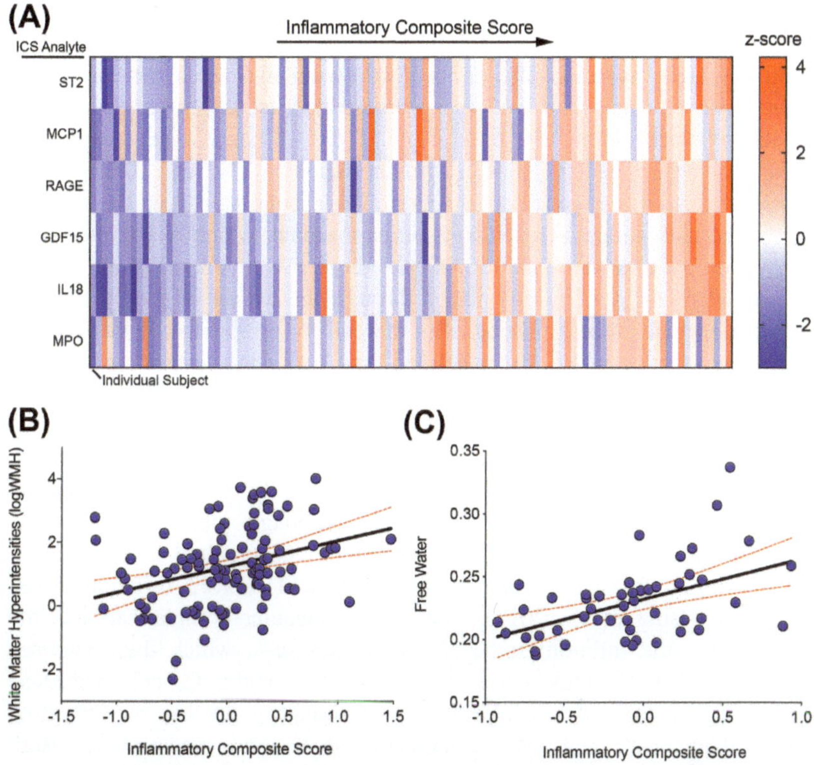

FIGURE 10.5 **ICS correlates with MRI measures of cerebrovascular injury.***ICS,* Heatmap of z scores for each of the individual analytes composing the ICS for each included subject in the MarkVCID cohort ordered left to right by ICS (average z-score of each analyte) (A). Scatter plot and regression line of logWMH versus ICS ($n = 110$) (B). Scatter plot and regression line of free water versus ICS ($n = 49$) (C). Red dashed lines indicate 95% confidence intervals. Inflammatory composite score; *MRI,* magnetic resonance imaging. *From Altendahl, M., Maillard, P., Harvey, D., Cotter, D., Walters, S., Wolf, A., Singh, B., Kakarla, V., Azizkhanian, I., Sheth, S. A., Xiao, G., Fox, E., You, M., Leng, M., Elashoff, D., Kramer, J. H., Decarli, C., Elahi, F. & Hinman, J. D. (2020). An IL-18-centered inflammatory network as a biomarker for cerebral white matter injury.* PloS One, 15*(1), e0227835. https://doi.org/10.1371/journal.pone.0227835.*

dynamic contrast-enhanced MRI and the PET scanners can show the activation of the microglia. A study was done with patients with vascular cognitive impairment and WM disease. Regions of BBB disruption were separate from those of microglia activation (Walsh et al., 2021).

Microglia participate in both injury and repair. The MI type of microglia serves normal brain function, including participating in repair. The M2 type participates in disruptive actions involved in acute and chronic inflammation. Cellular debris from various brain injuries activates microglia by stimulation of receptors for DAMPs; since pathogen-associated molecular patterns are not involved, this is referred to as sterile inflammation. Misfolded amyloid and tau molecules act as DAMPs to activate microglia to remove debris; this is an important role for the microglia since accumulation of cellular debris eventually kills the cells. The removal of waste involves formation of the inflammasome, which is a two-step process: the first step involves priming the cell by upregulating the expression of the inflammasome components NLRP3, caspase-1 and pro-IL-1β. Cytokines such as TNF and IL-1β lead to NF-κB activation and gene transcription. The second step involves the formation and activation of the NLRP3 inflammasome, which is activated by a wide variety of unrelated stimuli. As cells die, the components of the neuron are normally removed by autophagy through the action of the lysosomes. This is a normal process that benefits the cell by removing the debris. The removal system can be overwhelmed and activate destructive cellular mechanisms. One such destructive pathway involves the activation of an endosomal/lysosomal pathway. Amyloid degradation products cause the activation of Rab5-GTP, with the swelling of neuronal Rab5-endosomes. Acidified lysosomes facilitate the removal of the by-products of neuronal metabolism. However, as the lysosomal autophagy mechanism is overwhelmed by a reduction in acidity, endosomes accumulate in the neuron. This results in the activation of the NLRP1 and NLRP3 inflammasomes (Fig. 10.6) (Ising et al., 2019). The large molecular complex attracts the ASC specks: the inflammasome activates capsase-1, which then activates pro-IL-1β and IL-18. This begins the process of damaging the cell membrane by creating pores, and amplifying the inflammatory process by activating additional microglia and creating molecules that recruit macrophages. Both IL-1β and ASC can be released into extracellular space, perpetuating the injury process. The permeabilized cell loses electrolytes, swells, and undergoes apoptosis.

Loss of NLRP3 inflammasome function reduces tau hyperphosphorylation and aggregation by regulating tau kinases and phosphatases. Tau activated the NLRP3 inflammasome and intracerebral injection of fibrillar Aβ-containing brain homogenates induced tau pathology in an NLRP3-dependent manner. These data identify an important role of microglia and NLRP3 inflammasome activation in the pathogenesis of tauopathies and support the amyloid-cascade hypothesis in AD, demonstrating that neurofibrillary tangles develop downstream of Aβ-induced microglial activation. Inflammasomes are important in chronic inflammation and NLRP3 is a molecule that is central to the role of the inflammasome (Swanson et al., 2019). This is demonstrated by the role of inflammasomes in other diseases such as rheumatoid arthritis and diseases with abnormal innate immunity. Primary progressive multiple sclerosis is another disease of WM; there are high levels of mRNA for NLP3, suggesting a chronic form of inflammation is involved in that form of MS (Malhotra et al., 2020). By analogy, a similar process may happen in the WM injury in BD. Animal models of rheumatoid arthritis and multiple sclerosis can be improved by treatment with agents that block

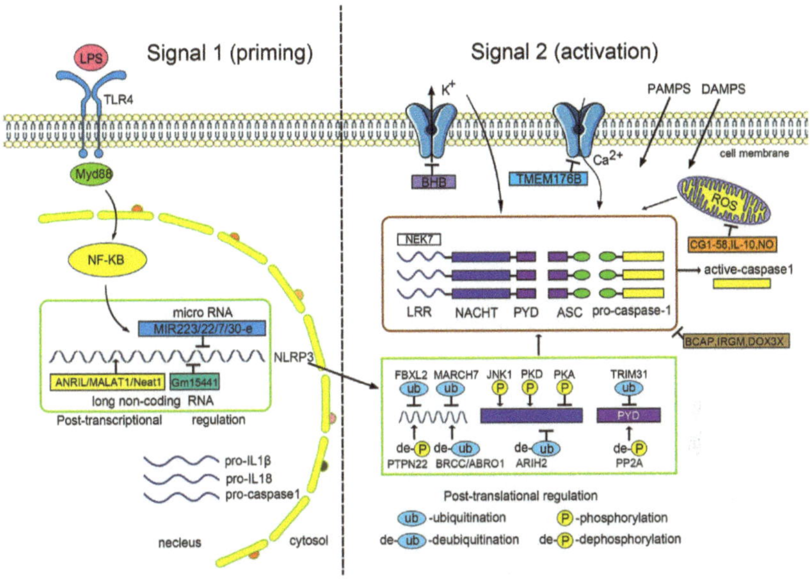

FIGURE 10.6 Activation of inflammasome. Activation of the NLRP3 inflammasome involves both the priming and activation steps. In the priming step (Signal 1), inflammatory triggers such as TLR4 agonist lipopolylsaccharide (LPS) induce the NF-κB-mediated expression of NLRP3, pro-IL-1β and procaspase-1. In the activation step, diverse PAMPs and DAMPs trigger the NLRP3 inflammasome assembly though unifying events such as ion flux, mitochondrial dysfunction and ROS formation. Regulation of the NLRP3 inflammasome activation can occur at the post-transcriptional and post-translational levels. MicroRNAs (e.g., MiR-223, MiR-22, MiR-7, and MiR-30e) inhibit the NLRP3 activity by targeting its untranslated region (UTR) binding sites, and long non-coding RNAs (e.g., ANRIL, MALAT1, Neat 1, and Gm15441) can either promote or inhibit the inflammasome signaling. The post-translational modifications at different sites and domains of NLRP3 protein include phosphorylation, dephosphorylation, ubiquitination, de-ubiquitination, and S-nitrosylation. Many molecules such as BCAP, IRGM, and DDX3X can interfere the assembly of NLRP3, ASC, and procaspase-1. Other negative regulatory molecules inhibit the inflammasome activation by targeting the K^+ efflux (e.g., BHB), cytosolic Ca^{2+} (e.g., TMEM176B), mitochondrial function (e.g., macrophage CGI-58, IL-10, and NO). *DAMPs*, Damage-associated molecular patterns; *NF-κB*, nuclear factor-κB; *PAMPs*, pathogen-associated molecular patterns. *From Zheng, D., Liwinski, T. & Elinav, E. (2020). Inflammasome activation and regulation: Toward a better understanding of complex mechanisms.* Cell Discovery, 6, *36. https://doi.org/10.1038/s41421-020-0167-x.*

the action of IL-1β or the inflammasome, suggesting that inhibiting the action of the inflammasome may prove to be a treatment for vascular dementia.

Senescent microglia and chronic inflammation

The microtubule-associated protein tau aggregates in multiple neurodegenerative diseases, causing inflammation and changing the inflammatory signature of microglia by unknown mechanisms. WM lesions gradually grow,

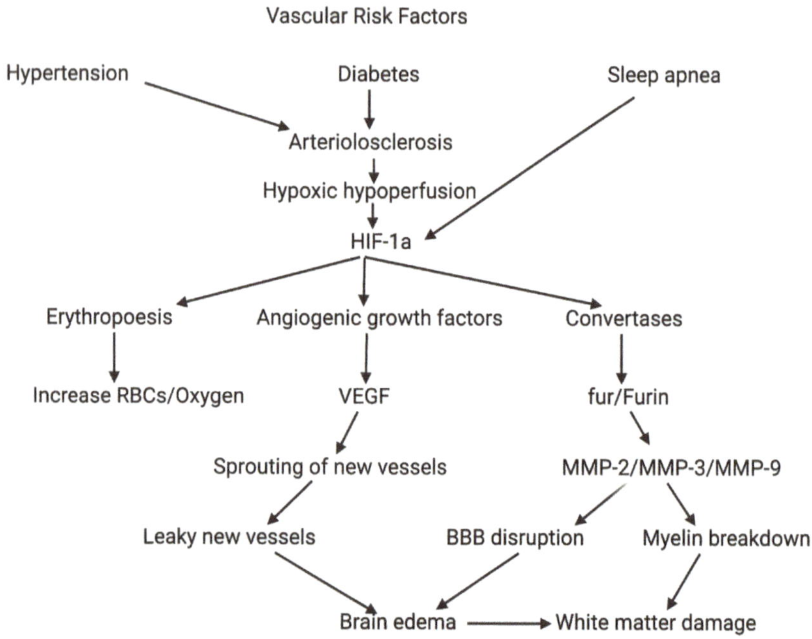

FIGURE 10.7 Schematic of the pathophysiology of white matter injury in vascular disease. The process begins with vascular risk factors, hypertension and diabetes, producing arteriolosclerosis and reducing cerebral blood flow to cause hypoxic hypoperfusion. Reduced oxygen delivery increases the expression of HIF-1α, activating a cascade of genes with production of proteins that attempt to restore oxygen, including erythropoiesis, angiogenesis, and activation proteases and cytokines. While increased RBC can be beneficial, and sprouting of vessels can increase blood flow, the newly formed vessels are leaky and increase brain edema. The increase in convertases, such as furin, causes the activation of the MMPs that breakdown the BBB, and release myelin basic protein and neurofilament light from the myelinated axons. Both the white matter edema and the loss of myelin result in the damage to the white matter that is seen on MRI. *BBB*, Blood–brain barrier; *HIF-1α*, hypoxia inducible factor-1α; *MMP*, matrix metalloproteinases; *MRI*, magnetic resonance imaging; *RBC*, red blood cells. *Figure made with BioRender.*

suggesting a chronic process. Microglia appear to play a major role in chronic inflammation. The multiple pathways initiated in inflammatory damage to the white matter are shown schematically in (Fig. 10.7). One possible mechanism involves the creation of senescent microglia that become a continuous source of proteases. When microglia phagocytose neurons containing tau aggregates, they induce the "eat me" signal of phosphatidylserine. The process is initiated when neurons phagocytize aggregated tau fragments. The neurons signal that they are going to die by expressing phosphatidylserine on the surface. This is a signal for the microglia to phagocytose the neuron. Microglia become hypophagocytic while releasing seed-competent insoluble tau aggregates. These microglia express a senescence-like phenotype, demonstrated by secretion of paracrine senescence-associated cytokines, and maturation of matrix remodeling enzymes, resulting in particular, the nuclear

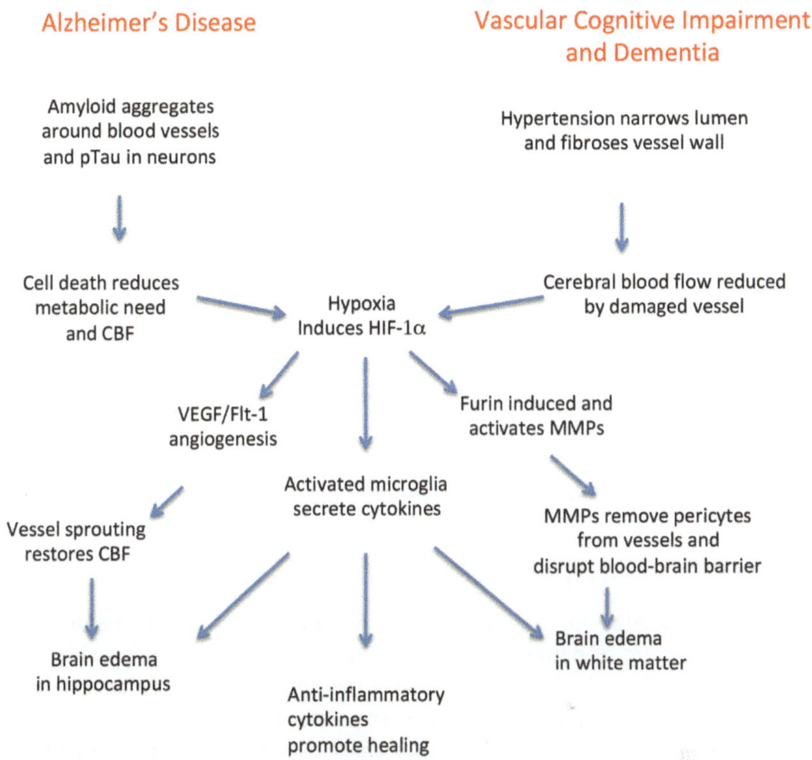

FIGURE 10.8 Schematic summary of molecular events involved leading to mixed dementia. Alzheimer's disease pathway on the left with amyloid and tau leading to cell death, reducing cerebral blood flow and inducing hypoxia inducible factor. On the right is the vascular disease pathway with hypertension damage to blood vessel initiating the further molecular events ending in brain edema and inflammation.

factor κB–dependent activation of matrix metalloprotease 3 (MMP-3/stromelysin1) (see Fig. 8.6).

After ingesting the dying neuron, the microglia enter a senescent state, which promotes the secretion of MMP-3. The MMP-3 is a toxic extracellular degrading protease that attacks the myelinated axons, setting up a situation where there is further tau incorporation into neurons, which are ingested by microglia. It is unclear what stops this feedback loop that creates a group of senescent microglia that contribute to the ongoing destruction of the axons. This process, which is called by-stander demyelination, is one of the pathological processes that attacks the blood vessels. This is a possible model for the Binswanger pathology that involves disruption of the BBB and breakdown of myelin (Brelstaff et al., 2021).

Inflammation is a key component of both amyloid/tau pathology and vascular pathology. Both vascular risk factors and amyloid proteins are involved in the inflammatory process in mixed dementia (Fig. 10.8). The senescent microglia are a logical link between the two processes that could explain the enhanced disease process seen in the dual or mixed pathology. Initially, amyloid buildup begins in some patients in their 40s, followed by the aggregation of tau inside the neurons, leading to their death. Expression of the apoptosis-inducing phosphatidylserine on the surface of the neuron. These "eat me" signals induce the microglia to phagocytose the ailing neurons. The induction of the cytokines and proteases attacks the myelinated fibers, extracellular matrix and tight junction proteins of the blood vessels. This could link the presence of Alzheimer's proteins in the brain with the augmentation of inflammation by the presence of vascular disease, which is another activator of microglia. The protease mediated by-stander demyelination was proposed earlier as the mechanism of inflammatory-based damage to the axons in MS (Cammer et al., 1978). This could be a mechanism for the cooperation of inflammation induced by amyloid and tau with that related to vascular disease. BD pathology is related to both the opening of the BBB and the breakdown of myelin, in which the senescent microglia could play a major role.

Insights into pathophysiology of Binswanger's disease

Treatment strategies for vascular cognitive impairment and dementia will depend on a group of patients with a natural history that can be reliably predicted, a pathophysiological model, and a treatment that will impact natural history. The term vascular cognitive impairment and dementia (VCID) encompasses all the forms of vascular disease that can impact cognition. Such a definition is too broad to be useful in clinical trials. By excluding forms of VCID without a clear natural history such as multiple strokes which are sporadic and do not necessarily lead to a defined pathway of progression, and single strategic strokes, which also do not necessarily progress, it leaves the forms of VCID involving small vessel disease with extensive involvement of the WM. There are two forms of small vessel disease, namely, lacunar strokes and BD. BD and lacunar state met three criteria: (1) vascular disease is the primary etiology; (2) ischemic WM lesions are the primary type of brain injury; and (3) subcortical location is the primary location of lesions (Erkinjuntti et al., 2000). This form of vascular dementia has been called SIVD. In a longitudinal study over 3 years, subcortical ischemic vascular disease (SIVD), as a manifestation of cerebral small vessel disease, was related to progressive cognitive impairment and a considerable risk of developing dementia. SIVD seems to specifically contribute to the deterioration of psychomotor speed, executive control, and global cognitive function (Jokinen et al., 2009). SIVD results from small vessel disease, which produces either arteriolar occlusion and lacunes or widespread incomplete infarction of WM due to stenosis of medullary arterioles and hypoperfusion, resulting in BD. Patients have executive dysfunction, dysarthria, apathy with mood changes, urinary incontinence, and imbalance. Brain imaging is necessary for correct diagnosis. Risk factors are advanced age, hypertension, diabetes, smoking, hyperhomocysteinemia, hyperfibrinogenemia, and other conditions that can cause brain hypoperfusion such as obstructive sleep apnea, congestive heart failure, cardiac arrhythmias, and orthostatic hypotension. Treatment is symptomatic and prevention requires control of treatable risk factors (Román et al., 2002).

Most patients with evidence of BD have other types of pathological changes at autopsy. The most common findings are the presence of misfolded proteins associated with AD (Toledo et al., 2013). A number of other types of misfolded proteins can be present at the same time; besides the abnormal proteins, vascular disease is a very common finding; generally, along with abnormal protein accumulations, there are found vascular diseases (Karanth et al., 2020; Schneider et al., 2007). Separating patients with different forms of pathology clinically is often impossible due to the overlapping of clinical symptomatology in the different forms of dementia. In addition, planning a rational approach to treatment and developing treatment trials is a major challenge without a relatively homogeneous group of patients with a known natural history. This was thought to be the case in patients with AD because of a progression of dementia that could be established with neuropsychological testing. However, using the approach that it is a disease based mainly on the presence of amyloid and tau has led to several large treatment trials to remove amyloid that have failed. One reason for the lack of positive results may have been the multiplicity of etiologies, and the treatment of one of them was insufficient to stem the progression.

Biomarkers that identify the underlying cause of the dementia at a molecular level have been developed. This biomarker-based approach has many benefits (Erhardt et al., 2021). Mainly, it provides insight into the underlying molecular events that are occurring in the brain. Early use of biomarkers was focused on CSF and PET amyloid and tau to identify those with AD; the biological approach to diagnosis has been recommended for use in AD research studies (Jack et al., 2018). The biological definition of AD is an important step toward the use of machine learning for diagnosis of dementia. Other causes of dementia can be added to the amyloid (A), phosphoTau (T), and neurodegeneration (N) formula to form a more complete picture of the underlying pathology in the brain. Vascular diseases are the most common comorbidity; the most important additional biomarker is to show the presence of vascular disease, since that is the most common comorbidity (Caprihan et al., 2021). The best biomarker to indicate vascular disease is injury to the WM as shown by diffusion tensor imaging (Raja et al., 2019). Adding a vascular factor to the ATN formula provides a means of showing four diagnostic categories: (1) pure AD with only CSF or PET evidence of Alzheimer proteins; (2) SIVD with abnormal diffusion tensor imaging (DTI) in the WM without Alzheimer proteins in CSF or PET, and (3) dual pathology of AD and vascular disease or MX. These are the three main forms of dementia. There is a fourth classification based on relatively normal cognition, but large WM lesions on FLAIR MRI. What to call these asymptomatic patients with MRIs that resemble those with BD has been challenging. One proposal is to call them leukoaraiosis, which was originally defined based on the radiological findings without ascribing pathological significance (Hachinski et al., 1987). Unfortunately, the definition of leukoaraisosis has been expanded beyond the intended use to include any WM changes. The use of biomarkers has been made more attractive with the introduction of ultrasensitive methods to detect in plasma small amounts of proteins from the brain, bypassing the need for an invasive lumbar puncture or an expensive PET scan. The availability of plasma biomarkers of amyloid and tau will enhance the ability to survey large populations of patients with signs of early cognitive impairment and to select those at an earlier stage of the disease. A further advance is the finding of measurable levels of NfL in plasma that could lead to a biomarker for the damage to the WM to augment the MRI.

Otto Binswanger's report of a vascular disease of WM associated with dementia was published in 1904, after a long period with relatively few publications to support his findings due to the need to make the diagnosis at an autopsy, which was rarely done in older persons until the development of neuroimaging in the 1970s spurred interest in pathological correlation with the MRIs. From being a rare disease, BD was frequently diagnosed based on radiological findings. By 1981, the editors of *Lancet* wrote that there was an epidemic of BD due to attenuation of the WM on computed tomography, warning against the use of an ancillary test as the basis of a diagnosis of dementia ("Binswanger's Encephalopathy," 1981). By the 1990s, BD had gone from a rare form of dementia primarily diagnosed at autopsy to one of the major causes of dementia, largely due to the demonstration of WM disease on MRI.

Advances in technology to study brain diseases since the 1970s has greatly improved the understanding of the pathophysiology of WM disease. The original observation of severe arteriosclerosis in the vessels of the WM with the resulting lack of nutrition, which was Binswanger's speculation as the cause of the disease, has been confirmed using studies of cerebral blood flow in humans. Studies in animals with hypertensive vascular disease and parallel studies in humans have verified the deficiency in cerebral blood flow that has affected the deep WM with its poor blood supply. This has been supported by animal studies that showed hypoxia in the WM in spontaneously hypertensive rats using paramagnetic resonance to directly measure the levels of oxygen in the WM over an extended period of time (Weaver et al., 2014). As a consequence of reduced oxygen, there is an increase in hypoxia inducible factor-1α (HIF-1α), initiating a cascade of molecular events to compensate for the low levels of oxygen. Erythropoiesis increases red blood cells to carry more oxygen, angiogenic growth factors are activated that lead to sprouting of new blood vessels from existing ones, and MMP are activated by the convertase, furin. Intermittent hypoxia is a stronger stimulus for HIF-1α than continuous loss of oxygen, which is the reason that sleep apnea is a potent risk factor for vascular cognitive impairment and dementia (Semenza, 2014).

There is a slow growth of the WM lesions that implies chronic progression; the growth is associated with decline in memory and slowing of gait over 3 years (Silbert et al., 2008). One possible cause of the progressive injury is the expression of MMP-3 by pericytes and senescent microglia. The role of the MMPs differs between acute ischemia where MMP-9 is the main inflammatory protease produced, and chronic hypoxia where MMP-3 is most prominent. In chronic WM injury, the CSF has MMP-2, MMP-3, and MMP-10, rather than MMP-9, which is seen in acute ischemia (Erhardt et al., 2021). The MMPs have two ways to damage the WM: (1) opening of the BBB in the WM, permitting toxins from the blood to enter the nervous system, and (2) a direct attack on the myelinated fibers (Chandler et al., 1995). Albumin index, the ratio of albumin in CSF to blood, is elevated in patients with vascular dementia. Albumin is produced in the liver and normally excluded from the brain by the BBB. Large molecules such as albumin enter the brain by pinocytosis, passing through the arterioles rather than the tight junctions (Nag et al., 1979). The vasogenic edema is seen best in the WM where it can flow along the WM fiber tracts as opposed to the densely packed neuropil in the gray matter, which restricts bulk flow. Molecules from the blood crossing the BBB, such as albumin, are toxic to WM (Hassel et al., 1994).

Microglia cells secrete MMP when they are activated; they also secrete inflammatory cytokines, including IL-6 and IL-8, TNF-α and IL-1β; these cytokines augment inflammation. In chronic conditions, such as atherosclerosis and rheumatoid arthritis, MMP-3 is the major protease producing the cellular injury. MMP-3 knockout mice have a blunted response to intracerebral injection of lipopolysaccharide (Gurney et al., 2006). Normally, microglia savage dead and dying cells to keep them from overwhelming the organism. Neurons that contain aggregated tau express phosphatidylserine on the cell surface, signaling microglia to ingest that cell. Once they have ingested the tau-containing neuron, the microglia enter a senescent state, and chronically produce MMP-3 (Brelstaff et al., 2021). This slow release of MMP-3 by senescent microglia, augmented by MMP-3 released by pericytes and activated microglia, may disrupt the BBB, allowing toxins from the blood to enter the brain; the presence of edema could account for the changes seen on diffusion tensor imaging, particularly in free water movement. DCE-MRI with gadolinium shows increased leakage in the normal appearing WM, which could account for the diffusion changes seen in WM (Huisa et al., 2015). Another action of MMP-3 is to degrade myelin in the axons, producing myelin basic protein and NfL, which can be measured in CSF and plasma. Thus, MMP-3 has the potential to create the milieu for the pathological changes seen in BD. Supporting the role of the MMPs in the pathology of BD is the presence of multiple MMPs in CSF of patients with dementia, particularly MMP-3 and MMP-10 (Erhardt et al., 2021). Minocycline, an inhibitor of MMPs, reduces WM damage and improves behavior in an animal model of vascular dementia, providing evidence to support the role of MMPs in pathophysiology (Jalal et al., 2015). Inhibitors of MMPs could interfere with the beneficial actions of MMPs, such as maintenance of synaptic function needed in learning. Specific MMP-9 inhibitors failed to block MMP injury in the hypertensive animal model, and they impaired learning most likely by interfering with synapse formation (Raz et al., 2018).

Another way to detect damage to the WM is to use myelin basic protein in the CSF or NfL in either CSF or blood. The presence of fragments of axons indicates that an inflammatory process has disrupted the fiber tracts. The ultrasensitive single molecule assays (SIMOA) is able to provide information on the presence of AD proteins, indicating a neurodegenerative process, and the breakdown of the myelinated axons. In the future, having this information will be sufficient to identify MX patients from a blood test. By avoiding the invasive lumbar puncture and the expensive PET scan as the methods to diagnose AD, SIVD and MX for clinical trials, it will be possible to screen large numbers of patients to find those at risk for one of the dementias. Theoretically, it should be possible to run the clinical trial at an earlier stage, improving the chances of success in smaller numbers of patients that have been identified by artificial intelligence in an era of precision medicine. Studies will be needed to compare MRI detection of WM damage with NfL blood tests to determine which is the most sensitive to detect the early changes in the WM. In either case, it will now be routinely possible to separate the mixed patients from those with pure AD and pure SIVD.

While CSF contains measurable levels of many of the analytes of interest, only a few can also be measured in plasma. Not only are plasma levels too low to detect, but there is the concern that they may be produced systemically rather than derived from the blood. Some proteins are made primarily by the brain, including phosphorylated tau and Aβ, and their levels in plasma are reliable indicators of brain production. Several of the MMPs are measurable in both the CSF and plasma, including MMP-3 and MMP-10. In addition, angiogenic growth factor VEFG-D (vascular endothelial growth factor-D) can be detected in both plasma and CSF. Cytokines cannot be measured in plasma and will require CSF (Erhardt et al., 2021). As the sensitivity and number of tests that can be done using plasma increases, it may be possible to avoid lumbar puncture and PET scanning in order to determine their levels in the brain. Classification and staging of patients should be possible entirely from plasma in the future; this would entail obtaining Aβ and phosphorylated tau to fulfill the AT part of the formula, and to use NfL as the V to show the presence of axonal damage, completing the classification. At this time, the separation of patients into the major groups of AD, MX, and SIVD is possible with plasma measurements of amyloid and tau with the addition of WM damage still needs to be shown by diffusion tensor MRI. The addition of biomarkers for myelin damage, such as NfL, may show WM injury, reducing the need for MRI. This blood-based approach to classification would facilitate screening of large populations of at-risk individuals, and would allow earlier diagnosis for initiation of more appropriate therapies based on the staging of the disease progression.

AD begins with the accumulation of metabolic waste products from amyloid that are found around blood vessels in the plaques; activated microglia and infiltrating macrophages are in the region. With time, abnormal forms of phosphorylated tau begin to be seen in neurons that will be removed by microglia. Eventually, the cell will undergo apoptosis and disappear. Slow flow of CSF out of the brain in the clogged perivascular pathways will enhance the buildup in the brain. These processes can begin in the 40s in patients with the genes for AD. On the other hand, vascular disease generally begins later in life, but in some with severe hypertensive syndromes, it can also begin as early in age as the 40s. The presence of vascular risk factors slowly affects the midsized arterioles; the lumen is narrowed, interfering with cerebral blood flow and the outer wall subjected to chronic increased pressure develops fibrosis, leading to poor response to physiological stimuli. The neurodegenerative and vascular processes progress slowly for many years, eventually crossing in later life, producing the MX that is the most prevalent form. Ideally, treatment would be individualized based on the presence of biomarkers in the plasma; treatments designed to remove amyloid and tau could be given at the first sign of an increase of their increase in plasma. Similarly, monitoring vascular risk factors from an early age would provide a guide as to when to initiate intensive treatment of vascular disease. Separate treatments would be necessary for the accumulation of misfolded proteins and for vascular disease. The pathophysiology of vascular disease is based on the presence of hypertension and diabetes, the two main vascular risk factors. Both are treatable and, if properly treated in midlife, could slow the onset of dementia in later life. Eventually, aging itself will overwhelm the system through a breakdown in the ability to fight inflammation. This process referred to as "inflammaging" will be a challenge to alter.

Will treatments aimed at reducing inflammation slow WM injury?

The treatment of dementia in the future will be based on results of biomarkers that show the pathophysiology by identifying the major pathological proteins affecting the brain. Those proteins are capable of triggering inflammation, which can become chronic and self-perpetuating. At what point in the inflammatory cascade will therapy be most effective? Blocking some stages may not be possible because the inflammation is too far advanced or the protein being inhibited may have multiple functions, some of which are beneficial. An example are the MMPs that have multiple roles: they are expressed during acute inflammation and exacerbate the injury process, which would be good to block. However, they are also important in consolidation of memory through their effect on synapse formation. This shows that treatment of inflammation during the time it is part of the repair process will have detrimental effects. Biomarkers may be able to determine the stage of the illness.

Will it be possible to impact inflammation by moving up the molecular cascade to the origin of the inflammation? An illustration of the pathways involved in the formation of inflammasomes is described earlier. Aging is a major factor in the onset of an inflammatory state. This may explain the tendency for inflammation to become more important as people age. Immunologic mechanisms have not been considered important in chronic neurodegeneration. This is changing with the information accumulating on the role of inflammasomes. At the center of these mechanisms is the activation of the NRLP3 complex, which initiates the production of IL-1β, IL-18 and TNF-α along with senescent microglia cells that perpetuate the chronicity of innate immunity. Modulation of inflammasome activity in models of AD, PD and other neurodegenerative diseases suggests a potential therapeutic role for substances that can effectively inhibit this immune signaling mechanism. Several large and small pharmaceutical companies are actively searching for small molecule inhibitors of inflammasomes, and several are in clinical trials.

Can biomarker-driven precision medicine be integrated into clinical practice?

Internal medicine has entered the era of precision medicine with the use of machine learning algorithms to make diagnostic and therapeutic decisions based on the use of "big data." Neurology has lagged in this effort, but with the advent of sophisticated diagnostic tools, such as MRI and the use of multiplex assays for blood and CSF analysis, neurology will soon catch up. As it enters the era of "big data," neurologists will need to learn to integrate the large amount of data available into the care of patients. Clinical data was refined in the 18th century and effectively combined with neuropathological data in the 19th century, but remained in a limited stage of development until neuroimaging, genetics, and biochemical studies expanded dramatically in the later part of the 20th century. Information obtained from the clinical examination and the patient's history can provide the major information to arrive at the correct diagnosis, particularly with well-trained clinicians. This was sufficient when there were few treatments for neurological diseases. As the number of diseases for which there are treatments increases, it has become imperative to be more precise in order to select a treatment tailored to the manifestations of each disease.

Precision medicine is most advanced in cancer treatment, where the genes can indicate the specific type of treatment to use. When a large amount of data is available, it is essential to have a way to analyze the data: computer-based machine learning coupled with artificial intelligence will enable the neurologist of the future to have access to large datasets and computer algorithms to mine them. This information will augment the clinical history and neurological examination with data from neuroimaging, biochemical and genetic studies. Machine learning incorporates experience to improve prediction and optimize future treatments. Machine learning incorporates training into the algorithms that will increase in accuracy as the information in the databases increases. Comparison with autopsy results will ultimately be necessary to confirm the prediction of brain pathology from the multimodal data.

In the evaluation of patients with dementia, neuropsychological testing will remain the essential feature to demonstrate the extent of cognitive impairment, and to localize abnormalities in specific areas of the brain. Differentiation of the major forms of dementia requires a careful assessment of cognition. In primarily cortical diseases, such as AD and frontotemporal dementia (FTLD), the type of cognitive impairment is critical for diagnosis. Evidence of motor involvement suggests the presence of Lewy body dementia and dementia in Parkinson's disease. It is challenging to identify patients with MCI that will convert to AD, and artificial intelligence has begun to be used to extract reliable predictors and to automatically classify different AD phenotypes. There was high level of heterogeneity observed in a meta-analysis to assess the contribution of machine learning and neuropsychological measures to the automated classification of MCI patients and the prediction of their conversion to AD, A meta-analysis showed that machine learning applied to neuropsychological measures can lead to a successful automatic classification, being more specific as a screening rather than a prognostic tool (Battista et al., 2020).

With the growing number of imaging modalities available to use in diagnosing dementia, complex computer analysis has become an important component of the inquiry. Structural data is extracted from the T1-weighted image by programs such as FreeSurfer that can measure all the regions of the brain with great precision. Atrophy of the cortex and the hippocampus can be extracted from the data and used in the diagnostic formulas being developed for AD; these gray matter regions also provide some information in patients with vascular cognitive impairment and dementia, but the diffusion-based methods are more important since they detect disease of the WM. Diffusion tensor imaging relies on computer analysis and goes beyond the visual inspection of the FLAIR image, which shows WM hyperintensities in normal individuals that should not be interpreted as involvement in dementia. Subtle changes in the DTI signal can detect myelin damage, water accumulation, and direction flow of the water; all are important in differentiating the underlying disease state. Changes in blood flow and oxygenation state provide the basis for the BOLD signal that can indicate regions of the brain that are activated during performance of a task or in the resting brain. Connections between brain regions can be found through highly computational correlation methods with accurate tracing of fiber tracts, including those that cross each other. With the advances in computational power, allowing the integration of structural and functional data, the ability to separate patients based on MRI has become possible.

The ultimate goal of identifying the underlying pathological processes that are occurring as brain function deteriorates and accomplishing this prior to death has begun to be achievable with the use of multiplexing protein assay systems that measure molecules in quantities as small as femtograms. The molecular fingerprints ultimately will be able to be defined in each of the major dementing diseases, but more importantly, will indicate the appropriate treatment. Then it will be critical to understand the evolution of these various pathological changes: some will be present early in the disease and others only at later stages. With this information, the stage of the disease will become known and the appropriate treatment used.

With this plethora of information available on growing numbers of patients, many of whom will have autopsy confirmation, a machine learning approach will be needed. The large datasets will form the basis for the training sets in which the diagnoses confirmed at autopsy are known. Comparing new patients with existing ones in the training sets, diagnoses will become more accurate. Finally, the effect of the treatments in the training sets will be able to predict the optimal treatments.

Classification of dementia patients can be facilitated using multiple biomarkers: neuropsychological testing, MRI/DTI, and CSF. This is the first step in a decision tree classification scheme. For example, patients with AD have memory loss, normal WM on MRI, and abnormal AD proteins in the CSF. Similar movement down the tree can be done with the other classes. Since this is the initial attempt at classification, it does not include the large number of biomarkers that can be obtained from multiplex assay arrays that test for proteases, angiogenic factors, and cytokines: these arrays of analytes indicate those undergoing BBB breakdown by proteases, angiogenesis in attempts at repair, and the presence of anti- and pro-inflammatory processes. Because of the large number of factors involved in the biochemical processes, it is necessary to use high level statistical analysis; at this stage machine learning is important, and the use of training sets imperative (Fig. 10.9).

The majority of current studies concentrate on the separation of different forms of AD to predict the progression from MCI to AD or the identification of dementias related to AD. This approach will become more important as biomarkers for the α-synuclein and TDP-43 proteins as well as others yet to be identified. However, few studies have attempted to differentiate between AD and VCID, and more importantly to find patients with MX, which is the most common form. The ability of neuropsychological tests and neuroimaging to make this distinction is beginning to be developed.

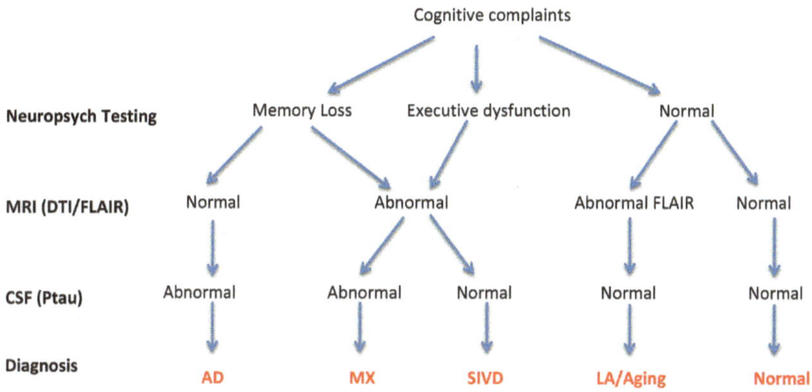

FIGURE 10.9 **Multimodal approach to patient classification.** Schematic of decision tree for patients with cognitive complaints. On the left are shown the testing performed, starting with neuropsychological testing for memory and executive performance; adding biomarker data from neuroimaging, including FLAIR and DTI; continuing with biochemical data from CSF to identify those with AD proteins. The end result is to bin patients into AD, MX, SIVD, LA/aging, and finally normal. *AD*, Alzheimer's disease; *CSF*, cerebrospinal fluid; *FLAIR*, fluid-attenuated inversion recovery; *LA*, leukoaraiosis; *MX*, mixed dementia; *SIVD*, subcortical ischemic vascular disease.

As the number of biomarkers increases and the pathological pathways involved in the major forms of dementia expands, the processes will be necessarily more complex, making them more difficult to model. A possible schema that can be used in the future involves pathways that start independently with either accumulation of proteins such as amyloid or vascular pathology, but at points converge and later diverge. Understanding the timing will be critical for making therapeutic decisions based both on the molecular pathways involved and the time in the disease process.

A number of misfolded proteins trigger neuroinflammation, including α-synuclein, TDP-43, amyloid, and tau. All of them can accumulate in the brain and aggregate to form toxic molecules with increasing age. Currently, only amyloid and tau can be measured during life with the others requiring an autopsy. α-Synuclein is a highly soluble unfolded protein that accumulates in Lewy bodies and Lewy bodies in Parkinson's disease. In the brain, α-synuclein is found mainly at the tips of neurons in presynaptic terminals. TDP-43 pathology, ranging from sparse to severe, is associated with amyloid plaques, tangles, and hippocampal sclerosis. After controlling for amyloid plaques, tangles, and hippocampal sclerosis, TDP-43 pathology was associated with more rapid cognitive decline and accounted for nearly as much of the variability in rates of global cognitive decline as did tangles. TDP-43 pathology had a distinct cognitive profile that differed from other neuropathologic processes (related to decline in episodic and working memory but not in other cognitive domains), and it was elevated in those who developed dementia but not in those with MCI, suggesting that TDP-43 is important in brain pathology, underlying cognitive decline and dementia in old age (Wilson et al., 2013).

While the clinical syndromes of FTLD are clinically and neuropathologically heterogeneous, neuroinflammation may be common across the disease spectrum. PET studies with a biomarker for activated microglia cells, ^{11}C-PK-11195, which is a proxy index of neuroinflammation, demonstrated the localization of tau and TDP-43 pathology, and the heterogeneity of clinical disease. There was a strong positive correlation between ^{11}C-PK-11195 and ^{18}F-AV-1451, which binds to sites in the brain associated with tau protein across widespread cortical regions. There was a strong association between the regional densities of microglia and neuropathology in FTLD. This was driven by ameboid (activated) microglia, with no change in the density of ramified (sessile) microglia. These in vivo findings indicate a close association between neuroinflammation and protein aggregation in FTLD and that the inflammatory component may be important in shaping the clinical and neuropathological patterns of the diverse clinical syndromes (Bevan-Jones et al., 2020).

All of these abnormal accumulations of proteins in different parts of the nervous system appear to contribute to the inflammatory response in the brain. Separating the contribution of each to the total picture will be difficult, but most likely the presence of multiple inciting factors will accelerate the clinical decline in both cognition and motor function. Vascular disease is a major cause of inflammation due to the damage to the blood vessels and a hypoxic environment in the brain. It is well known that the combination of amyloid pathology and ischemic disease is present in a high percentage of patients with dementia and hastens the disease process.

Neurology training in the future

The neurologists of the 21st century have come a long way in a short time from the descriptive neurology that was a mainstay of clinical training as late as the 1950s. It was based on neurological findings obtained from an elaborate examination. We are in a transition period as the field of neurology moves from diagnoses based on clinical history and examination to the use of modern imaging methods and biochemical analysis of blood and CSF that are currently available. Less and less time is now spent on doing a meticulous neurological examination to precisely localize the site of the problem in the nervous system. The importance of knowing the exact location of the problem will remain a challenge for the neurologist, but as the sophistication and precision of neuroimaging and biochemical studies continues to grow, there will be greater reliance on the discovery of the multimodal biomarkers that reveal the underlying pathology in the brain, which will ultimately be the guide to the new biological therapies that are now being developed. Although this may sound like science fiction, it is already happening in other fields of medicine, such as oncology, where the computer coupled with artificial intelligence algorithms is providing support to the clinician. Some clinicians will be able to achieve a high level of diagnostic acumen through diligent practice coupled with highly refined observational powers, but the decisions regarding appropriate treatment at each stage of the illness will be dependent on the ability to collect data from neuropsychological testing, neuroimaging, and biochemical studies. It will be unlikely that one individual will master all these fields in such a way as to be able to integrate the information into a useful form. The patient encounters to obtain history and neurological examination will remain the work of the clinician. Access to advanced imaging and laboratories capable of analyzing blood and CSF, and the ability to perform high level neuroimaging, will become routine in the teams that will be formed in the future. At this time, we do not know how effective a precision machine approach will be, particularly since the precise information on the underlying pathology cannot yet be translated into an effective treatment. The early neurologists made great strides with limited resources, using exceptional observational skills. They had great patience as they awaited the availability of the pathological data to correlate with the earlier clinical course. Despite the need for great integrative skills, many of the diseases slowly revealed themselves. The challenge of emerging neurologists will be to incorporate all this prior knowledge into a framework that includes a large amount of ancillary data, some of which will be more important than others.

References

Altendahl, M., Maillard, P., Harvey, D., Cotter, D., Walters, S., Wolf, A., Singh, B., Kakarla, V., Azizkhanian, I., Sheth, S. A., Xiao, G., Fox, E., You, M., Leng, M., Elashoff, D., Kramer, J. H., Decarli, C., Elahi, F., & Hinman, J. D. (2020). An IL-18-centered inflammatory network as a biomarker for cerebral white matter injury. *PLoS One*, *15*(1), e0227835. Available from https://doi.org/10.1371/journal.pone.0227835.

Anthony, D. C., Ferguson, B., Matyzak, M. K., Miller, K. M., Esiri, M. M., & Perry, V. H. (1997). Differential matrix metalloproteinase expression in cases of multiple sclerosis and stroke. *Neuropathology and Applied Neurobiology*, *23*(5), 406–415.

Barthel, H., Gertz, H.-J., Dresel, S., Peters, O., Bartenstein, P., Buerger, K., Hiemeyer, F., Wittemer-Rump, S. M., Seibyl, J., Reininger, C., & Sabri, O. (2011). Cerebral amyloid-β PET with florbetaben (18F) in patients with Alzheimer's disease and healthy controls: A multicentre phase 2 diagnostic study. *The Lancet Neurology*, *10*(5), 424–435. Available from https://doi.org/10.1016/S1474-4422(11)70077-1.

Battista, P., Salvatore, C., Berlingeri, M., Cerasa, A., & Castiglioni, I. (2020). Artificial intelligence and neuropsychological measures: The case of Alzheimer's disease. *Neuroscience and Biobehavioral Reviews*, *114*, 211–228. Available from https://doi.org/10.1016/j.neubiorev.2020.04.026.

Bevan-Jones, W. R., Cope, T. E., Jones, P. S., Kaalund, S. S., Passamonti, L., Allinson, K., Green, O., Hong, Y. T., Fryer, T. D., Arnold, R., Coles, J. P., Aigbirhio, F. I., Larner, A. J., Patterson, K., O'Brien, J. T., & Rowe, J. B. (2020). Neuroinflammation and protein aggregation co-localize across the frontotemporal dementia spectrum. *Brain: A Journal of Neurology*, *143*(3), 1010–1026. Available from https://doi.org/10.1093/brain/awaa033.

Fisher, C. M. (1981). Binswanger's encephalopathy. *Lancet (London, England)*, *1*(8226), 923.

Biswal, B., Yetkin, F. Z., Haughton, V. M., & Hyde, J. S. (1995). Functional connectivity in the motor cortex of resting human brain using echo-planar MRI. *Magnetic Resonance in Medicine*, *34*(4), 537–541.

Brelstaff, J. H., Mason, M., Katsinelos, T., McEwan, W. A., Ghetti, B., Tolkovsky, A. M., & Spillantini, M. G. (2021). Microglia become hypofunctional and release metalloproteases and tau seeds when phagocytosing live neurons with P301S tau aggregates. *Science Advances*, *7*(43), eabg4980. Available from https://doi.org/10.1126/sciadv.abg4980.

Brown, R. B., Traylor, M., Burgess, S., Sawcer, S., & Markus, H. S. (2019). Do cerebral small vessel disease and multiple sclerosis share common mechanisms of white matter injury? *Stroke: A Journal of Cerebral Circulation*, *50*(8), 1968–1972. Available from https://doi.org/10.1161/STROKEAHA.118.023649.

Cammer, W., Bloom, B. R., Norton, W. T., & Gordon, S. (1978). Degradation of basic protein in myelin by neutral proteases secreted by stimulated macrophages: A possible mechanism of inflammatory demyelination. *Proceedings of the National Academy of Sciences of the United States of America*, *75*(3), 1554–1558.

Caprihan, A., Raja, R., Hillmer, L. J., Erhardt, E. B., Prestopnik, J., Thompson, J., Adair, J. C., Knoefel, J. E., & Rosenberg, G. A. (2021). A double-dichotomy clustering of dual pathology dementia patients. *Cerebral Circulation – Cognition and Behavior*, *2*. Available from https://doi.org/10.1016/j.cccb.2021.100011.

Chandler, S., Coates, R., Gearing, A., Lury, J., Wells, G., & Bone, E. (1995). Matrix metalloproteinases degrade myelin basic protein. *Neuroscience Letters*, *201*(3), 223–226.

Erhardt, E. B., Adair, J. C., Knoefel, J. E., Caprihan, A., Prestopnik, J., Thompson, J., Hobson, S., Siegel, D., & Rosenberg, G. A. (2021). Inflammatory biomarkers aid in diagnosis of dementia. *Frontiers in Aging Neuroscience*, *13*, 717344. Available from https://doi.org/10.3389/fnagi.2021.717344.

Erkinjuntti, T., Inzitari, D., Pantoni, L., Wallin, A., Scheltens, P., Rockwood, K., Roman, G. C., Chui, H., & Desmond, D. W. (2000). Research criteria for subcortical vascular dementia in clinical trials. *Journal of Neural Transmission. Supplementum*, *59*, 23–30.

Fu, Z., Caprihan, A., Chen, J., Du, Y., Adair, J. C., Sui, J., Rosenberg, G. A., & Calhoun, V. D. (2019). Altered static and dynamic functional network connectivity in Alzheimer's disease and subcortical ischemic vascular disease: Shared and specific brain connectivity abnormalities. *Human Brain Mapping*, *40*(11), 3203–3221. Available from https://doi.org/10.1002/hbm.24591.

Fu, Z., Iraji, A., Caprihan, A., Adair, J. C., Sui, J., Rosenberg, G. A., & Calhoun, V. D. (2020). In search of multimodal brain alterations in Alzheimer's and Binswanger's disease. *NeuroImage Clinical*, *26*, 101937. Available from https://doi.org/10.1016/j.nicl.2019.101937.

Gurney, K. J., Estrada, E. Y., & Rosenberg, G. A. (2006). Blood-brain barrier disruption by stromelysin-1 facilitates neutrophil infiltration in neuroinflammation. *Neurobiology of Disease*, *23*(1), 87–96.

Hachinski, V. C., Potter, P., & Merskey, H. (1987). Leuko-araiosis. *Archives of Neurology*, *44* (1), 21–23.

Hassel, B., Iversen, E. G., & Fonnum, F. (1994). Neurotoxicity of albumin in vivo. *Neuroscience Letters*, *167*(1–2), 29–32.

Huisa, B. N., Caprihan, A., Thompson, J., Prestopnik, J., Qualls, C. R., & Rosenberg, G. A. (2015). Long-term blood-brain barrier permeability changes in Binswanger disease. *Stroke: A Journal of Cerebral Circulation*, *46*(9), 2413–2418. Available from https://doi.org/10.1161/STROKEAHA.115.009589.

Ising, C., Venegas, C., Zhang, S., Scheiblich, H., Schmidt, S. V., Vieira-Saecker, A., Schwartz, S., Albasset, S., McManus, R. M., Tejera, D., Griep, A., Santarelli, F., Brosseron, F., Opitz,

S., Stunden, J., Merten, M., Kayed, R., Golenbock, D. T., Blum, D., ... Heneka, M. T. (2019). NLRP3 inflammasome activation drives tau pathology. *Nature*, *575*(7784), 669–673. Available from https://doi.org/10.1038/s41586-019-1769-z.

Jack, C. R., Bennett, D. A., Blennow, K., Carrillo, M. C., Dunn, B., Haeberlein, S. B., Holtzman, D. M., Jagust, W., Jessen, F., Karlawish, J., Liu, E., Molinuevo, J. L., Montine, T., Phelps, C., Rankin, K. P., Rowe, C. C., Scheltens, P., Siemers, E., Snyder, H. M., & Sperling, R. (2018). NIA-AA research framework: Toward a biological definition of Alzheimer's disease. *Alzheimer's & Dementia: The Journal of the Alzheimer's Association*, *14*(4), 535–562. Available from https://doi.org/10.1016/j.jalz.2018.02.018.

Jalal, F. Y., Yang, Y., Thompson, J. F., Roitbak, T., & Rosenberg, G. A. (2015). Hypoxia-induced neuroinflammatory white-matter injury reduced by minocycline in SHR/SP. *Journal of Cerebral Blood Flow and Metabolism : Official Journal of the International Society of Cerebral Blood Flow and Metabolism*, *35*(7), 1145–1153. Available from https://doi.org/10.1038/jcbfm.2015.21.

Jokinen, H., Kalska, H., Ylikoski, R., Madureira, S., Verdelho, A., van der Flier, W. M., Scheltens, P., Barkhof, F., Visser, M. C., Fazekas, F., Schmidt, R., O'Brien, J., Waldemar, G., Wallin, A., Chabriat, H., Pantoni, L., Inzitari, D., & Erkinjuntti, T. (2009). Longitudinal cognitive decline in subcortical ischemic vascular disease–the LADIS Study. *Cerebrovascular Diseases (Basel, Switzerland)*, *27*(4), 384–391. Available from https://doi.org/10.1159/000207442.

Jonsson, M., Zetterberg, H., van Straaten, E., Lind, K., Syversen, S., Edman, A., Blennow, K., Rosengren, L., Pantoni, L., Inzitari, D., & Wallin, A. (2010). Cerebrospinal fluid biomarkers of white matter lesions – cross-sectional results from the LADIS study. *European Journal of Neurology*, *17*(3), 377–382. Available from https://doi.org/10.1111/j.1468-1331.2009.02808.x.

Karanth, S., Nelson, P. T., Katsumata, Y., Kryscio, R. J., Schmitt, F. A., Fardo, D. W., Cykowski, M. D., Jicha, G. A., Van Eldik, L. J., & Abner, E. L. (2020). Prevalence and clinical phenotype of quadruple misfolded proteins in older adults. *JAMA Neurology*, *77*(10), 1299–1307. Available from https://doi.org/10.1001/jamaneurol.2020.1741.

Khalil, M., Teunissen, C. E., Otto, M., Piehl, F., Sormani, M. P., Gattringer, T., Barro, C., Kappos, L., Comabella, M., Fazekas, F., Petzold, A., Blennow, K., Zetterberg, H., & Kuhle, J. (2018). Neurofilaments as biomarkers in neurological disorders. *Nature Reviews Neurology*, *14*(10), 577–589. Available from https://doi.org/10.1038/s41582-018-0058-z.

Klunk, W. E., Engler, H., Nordberg, A., Wang, Y., Blomqvist, G., Holt, D. P., Bergström, M., Savitcheva, I., Huang, G., Estrada, S., Ausén, B., Debnath, M. L., Barletta, J., Price, J. C., Sandell, J., Lopresti, B. J., Wall, A., Koivisto, P., Antoni, G., ... Långström, B. (2004). Imaging brain amyloid in Alzheimer's disease with Pittsburgh compound-B. *Annals of Neurology*, *55*(3), 306–319.

Kok, E., Haikonen, S., Luoto, T., Huhtala, H., Goebeler, S., Haapasalo, H., & Karhunen, P. J. (2009). Apolipoprotein E-dependent accumulation of Alzheimer disease-related lesions begins in middle age. *Annals of Neurology*, *65*(6), 650–657. Available from https://doi.org/10.1002/ana.21696.

Liu, C.-C., Liu, C.-C., Kanekiyo, T., Xu, H., & Bu, G. (2013). Apolipoprotein E and Alzheimer disease: Risk, mechanisms and therapy. *Nature Reviews Neurology*, *9*(2), 106–118. Available from https://doi.org/10.1038/nrneurol.2012.263.

Lv, H., Wang, Z., Tong, E., Williams, L. M., Zaharchuk, G., Zeineh, M., Goldstein-Piekarski, A. N., Ball, T. M., Liao, C., & Wintermark, M. (2018). Resting-state functional MRI: Everything that nonexperts have always wanted to know. *AJNR. American Journal of Neuroradiology*, *39*(8), 1390–1399. Available from https://doi.org/10.3174/ajnr.A5527.

Malhotra, S., Costa, C., Eixarch, H., Keller, C. W., Amman, L., Martínez-Banaclocha, H., Midaglia, L., Sarró, E., Machín-Díaz, I., Villar, L. M., Triviño, J. C., Oliver-Martos, B., Parladé, L. N., Calvo-Barreiro, L., Matesanz, F., Vandenbroeck, K., Urcelay, E., Martínez-Ginés, M.-L., Tejeda-Velarde, A., … Comabella, M. (2020). NLRP3 inflammasome as prognostic factor and therapeutic target in primary progressive multiple sclerosis patients. *Brain: A Journal of Neurology*, *143*(5), 1414–1430. Available from https://doi.org/10.1093/brain/awaa084.

Nag, S., Robertson, D. M., & Dinsdale, H. B. (1979). Quantitative estimate of pinocytosis in experimental acute hypertension. *Acta Neuropathologica*, *46*(1–2), 107–116.

Pascoal, T. A., Benedet, A. L., Ashton, N. J., Kang, M. S., Therriault, J., Chamoun, M., Savard, M., Lussier, F. Z., Tissot, C., Karikari, T. K., Ottoy, J., Mathotaarachchi, S., Stevenson, J., Massarweh, G., Schöll, M., de Leon, M. J., Soucy, J.-P., Edison, P., Blennow, K., … Rosa-Neto, P. (2021). Microglial activation and tau propagate jointly across Braak stages. *Nature Medicine*, *27*(9), 1592–1599. Available from https://doi.org/10.1038/s41591-021-01456-w.

Raja, R., Rosenberg, G., & Caprihan, A. (2019). Review of diffusion MRI studies in chronic white matter diseases. *Neuroscience Letters*, *694*, 198–207. Available from https://doi.org/10.1016/j.neulet.2018.12.007.

Raz, L., Yang, Y., Thompson, J., Hobson, S., Pesko, J., Mobashery, S., Chang, M., & Rosenberg, G. (2018). MMP-9 inhibitors impair learning in spontaneously hypertensive rats. *PLoS One*, *13*(12), e0208357. Available from https://doi.org/10.1371/journal.pone.0208357.

Román, G. C., Erkinjuntti, T., Wallin, A., Pantoni, L., & Chui, H. C. (2002). Subcortical ischaemic vascular dementia. *The Lancet Neurology*, *1*(7), 426–436.

Schneider, J. A., Arvanitakis, Z., Bang, W., & Bennett, D. A. (2007). Mixed brain pathologies account for most dementia cases in community-dwelling older persons. *Neurology*, *69*(24), 2197–2204.

Semenza, G. L. (2014). Oxygen sensing, hypoxia-inducible factors, and disease pathophysiology. *Annual Review of Pathology*, *9*, 47–71. Available from https://doi.org/10.1146/annurev-pathol-012513-104720.

Silbert, L. C., Nelson, C., Howieson, D. B., Moore, M. M., & Kaye, J. A. (2008). Impact of white matter hyperintensity volume progression on rate of cognitive and motor decline. *Neurology*, *71*(2), 108–113. Available from https://doi.org/10.1212/01.wnl.0000316799.86917.37.

Swanson, K. V., Deng, M., & Ting, J. P.-Y. (2019). The NLRP3 inflammasome: Molecular activation and regulation to therapeutics. *Nature Reviews Immunology*, *19*(8), 477–489. Available from https://doi.org/10.1038/s41577-019-0165-0.

Thijssen, E. H., La Joie, R., Wolf, A., Strom, A., Wang, P., Iaccarino, L., Bourakova, V., Cobigo, Y., Heuer, H., Spina, S., VandeVrede, L., Chai, X., Proctor, N. K., Airey, D. C., Shcherbinin, S., Duggan Evans, C., Sims, J. R., Zetterberg, H., Blennow, K., … Boxer, A. L. (2020). Diagnostic value of plasma phosphorylated tau181 in Alzheimer's disease and frontotemporal lobar degeneration. *Nature Medicine*, *26*(3), 387–397. Available from https://doi.org/10.1038/s41591-020-0762-2.

Toledo, J. B., Arnold, S. E., Raible, K., Brettschneider, J., Xie, S. X., Grossman, M., Monsell, S. E., Kukull, W. A., & Trojanowski, J. Q. (2013). Contribution of cerebrovascular disease in autopsy confirmed neurodegenerative disease cases in the National Alzheimer's Coordinating Centre. *Brain: A Journal of Neurology*, *136*(Pt 9), 2697–2706. Available from https://doi.org/10.1093/brain/awt188.

Walsh, J., Tozer, D. J., Sari, H., Hong, Y. T., Drazyk, A., Williams, G., Shah, N. J., O'Brien, J. T., Aigbirhio, F. I., Rosenberg, G., Fryer, T. D., & Markus, H. S. (2021). Microglial activation and blood-brain barrier permeability in cerebral small vessel disease. *Brain: A Journal of Neurology*, *144*(5), 1361–1371. Available from https://doi.org/10.1093/brain/awab003.

Weaver, J., Jalal, F. Y., Yang, Y., Thompson, J., Rosenberg, G. A., & Liu, K. J. (2014). Tissue oxygen is reduced in white matter of spontaneously hypertensive-stroke prone rats: A longitudinal study with electron paramagnetic resonance. *Journal of Cerebral Blood Flow and Metabolism: Official Journal of the International Society of Cerebral Blood Flow and Metabolism*, *34*(5), 890–896. Available from https://doi.org/10.1038/jcbfm.2014.35.

Wilson, R. S., Yu, L., Trojanowski, J. Q., Chen, E.-Y., Boyle, P. A., Bennett, D. A., & Schneider, J. A. (2013). TDP-43 pathology, cognitive decline, and dementia in old age. *JAMA Neurology*, *70*(11), 1418–1424. Available from https://doi.org/10.1001/jamaneurol.2013.3961.

Zetterberg, H., Skillbäck, T., Mattsson, N., Trojanowski, J. Q., Portelius, E., Shaw, L. M., Weiner, M. W., & Blennow, K. (2016). Association of cerebrospinal fluid neurofilament light concentration with Alzheimer disease progression. *JAMA Neurology*, *73*(1), 60–67. Available from https://doi.org/10.1001/jamaneurol.2015.3037.

Index

Note: Page numbers followed by "*f*" and "*t*" refer to figures and tables, respectively.

W

Kolibri 360 GmbH
Ferdinand-Jühlke-Straße 7
99095 Erfurt, Deutschland
produktsicherheit@kolibri360.de